HOW
MATH
WORKS

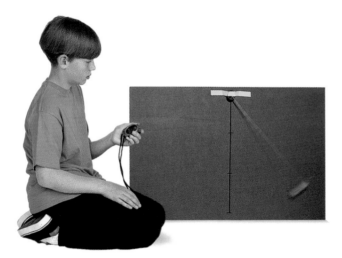

Using mental arithmetic to
play a spiral board game

Making music to show
fractional differences

Making a "magic wallet"
from two rectangles

Working out the volume of a hand

Finding the angle of
elevation with an astrolabe

Enlarging a small area

HOW
MATH
WORKS

Carol Vorderman

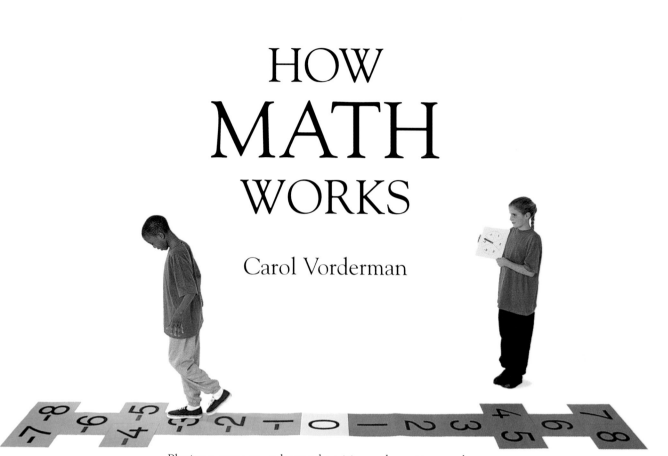

Playing a game to understand positive and negative numbers

Working out a formula
with a function machine

Juggling balls to form a
natural parabolic curve

Reader's Digest

The Reader's Digest Association, Inc.
Pleasantville, New York • Montreal

A READER'S DIGEST BOOK
designed and edited by Dorling Kindersley Limited, London

Project Editor	Charyn Jones
Art Editor	Gurinder Purewall
Editor	Katriona John
Designer	Elaine Monaghan
Production Controller	Louise Daly
Managing Art Editor	Nigel Duffield
Senior Managing Editor	Krystyna Mayer
Senior Managing Art Editor	Lynne Brown
Educational consultants	Roger Bridgman
	Jack Challoner
	Graham Hill
	Jim Miller
North American consultant	A. T. McPhee

The credits and acknowledgments that appear on page 192 are hereby made a part of this copyright page.

Copyright © 1996 Dorling Kindersley Limited, London

Text copyright © 1996 Carol Vorderman

Library of Congress Cataloging in Publication Data

Vorderman, Carol
    How math works / Carol Vorderman.
        p.   cm.
    Includes index.
    ISBN 0-89577-850-5
    1. Mathematics—Popular works.   I. Title.
QA93. V65   1996
510—dc20   95-47867

Printed in China
Fifth Printing, July 2001

Contents

Numbers

Proportions

Algebra

Statistics

Measurement

Shape

Thinking

INTRODUCTION

THE HISTORY OF MATHEMATICS runs from the ancient Egyptians, to the classical Greek scholars and natural philosophers, to Sir Isaac Newton and calculus, to today, where the purest mathematical study has its application in the exciting world of artificial intelligence and the transfer of digital information.

But math has set the course of history, too. For example, the ancient Egyptians invented a formal means of measuring using the "cubit," marking the measures out on a piece of black granite against which all the measuring sticks in the country had to be compared. Without this cubit, the great Pyramids of Gizeh may not have been built. In ancient Greece, Pythagoras and his school of disciples pointed out the harmony of music and mathematics, and formulated many properties of numbers and shapes. But they did so at their own cost: Pythagoras and many of his followers were murdered because of their studies.

During the Renaissance some of the greatest mathematicians were also artists. Leonardo da Vinci wrote with his left hand, in front of a mirror, so that his prose on various mathematical concepts could not be easily stolen. At the same time, mathematics enabled astronomers to challenge the view of people at the

time that the Earth was flat, and to suggest that the Earth was not the center of the Universe — a heresy punishable by death.

Just as literature cannot function without words, so science has no meaning without numbers and mathematical skills. In fact, math is as basic a need as words and communication. Without it our civilized world would fall apart.

If math did not exist there would be no buildings or bridges, no computers or telecommunications, no modern medicine or surgery, no cars or spacecraft, no financial or business worlds. Mathematics is not just about numbers and the logic of reasoning. It also helps us to understand the shapes of atoms and planets, to construct silicon chips and suspension bridges, and to analyze how and why a virus spreads.

How Math Works shows you where to look for math in your world — in the patterns on the heads of sunflowers, in the probability of winning a fortune, and in the stability of molecules according to their geometric structure. It reveals, too, the secrets of famous mathematicians through lively experiments that you can do yourself. Although you may never become a professional mathematician, the knowledge and excitement that you can gain on these pages will be useful to you throughout your life.

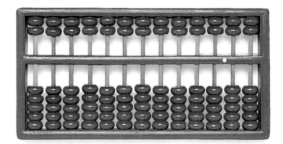

The home laboratory

ALTHOUGH MATHEMATICS uses some special technical equipment that can be bought in kits for school, most of the experiments in this book can be performed using household items, colored paper, and poster board. As well as your mathematics kit, there are items from the kitchen and the toolbox that can be used. On these two pages are shown some of the things that you will find in the lists of ingredients for the experiments.

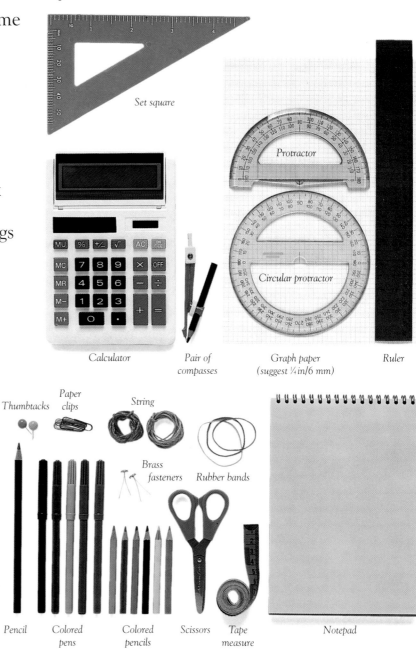

Set square

Calculator

Pair of compasses

Protractor

Circular protractor

Graph paper (suggest ¼ in/6 mm)

Ruler

■ Basic equipment

These items, some of which you may already have, are used in many of the experiments. The mathematics equipment (top right) is often used in school work. You need a ruler to draw and measure lines, and a set square for making and checking right angles. A pair of compasses enables you to draw accurate circles, and is useful for piercing or scoring. As well as an ordinary protractor, you need a circular one for measuring angles greater than 180°. Graph paper is used for drawing grids and graphs. You will often have to do calculations on a calculator. Make sure that yours has percent (%) and square root (√) keys. You should also put together some basic equipment, including pens and pencils, a notepad for writing down numbers and calculations, scissors, a tape measure, and string and rubber bands for tying and securing objects.

Thumbtacks *Paper clips* *String*

Brass fasteners *Rubber bands*

Pencil *Colored pens* *Colored pencils* *Scissors* *Tape measure* *Notepad*

■ Containers

Your home may already have items like the ones shown below. Wash them thoroughly after use; if they have held substances such as rubbing alcohol, you may not be able to use them for food and drink afterward. Some items are needed for measuring; a glass is good for nonspecific amounts, but use a measuring jug or cup for more accurate work. Build up a supply of empty objects, such as plastic bottles and toilet-roll inner tubes, for your experiments.

Glass bowl

Metal pail

Glass

Jug

Plastic bottle

Materials

Many experiments involve drawing or making things, so you will need to buy artists' materials from a stationer or art supply store. Some of these, such as masking tape and foamcore, may be unfamiliar to you, but they are inexpensive to buy and simple to use. Foamcore is available from specialist stationers. When buying glue, it is best to choose the type that comes as a stick, because it can be applied easily with no waste or mess.

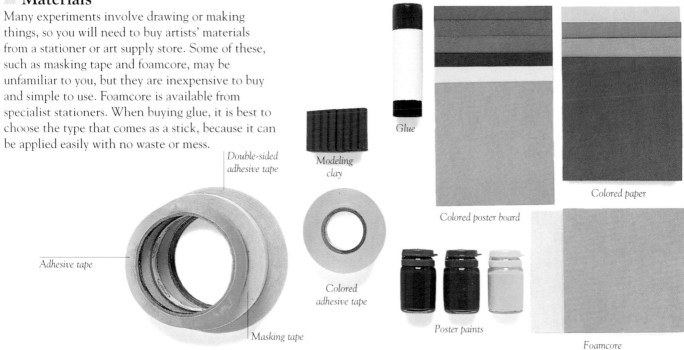

Double-sided adhesive tape

Modeling clay

Glue

Colored paper

Colored poster board

Adhesive tape

Colored adhesive tape

Poster paints

Foamcore

Masking tape

Useful items

Look in your kitchen for food coloring, straws, spoons, bamboo skewers, and mixing sticks for drinks. You may also find more specialized items there, such as a cooking thermometer and weighing scales. Use a stopwatch that shows fractions of a second; some digital wristwatches can double as stopwatches. Choose an ordinary thermometer that shows negative temperatures.

Board games often contain dice and counters, and you may have wooden balls, beads, and marbles among your belongings. To find wooden strips and dowels, look in the garage, basement, or garden shed, or ask for spare pieces from someone doing some decorating. Ask an adult to find a craft knife and help you use it. Spools of thread may be found in sewing equipment.

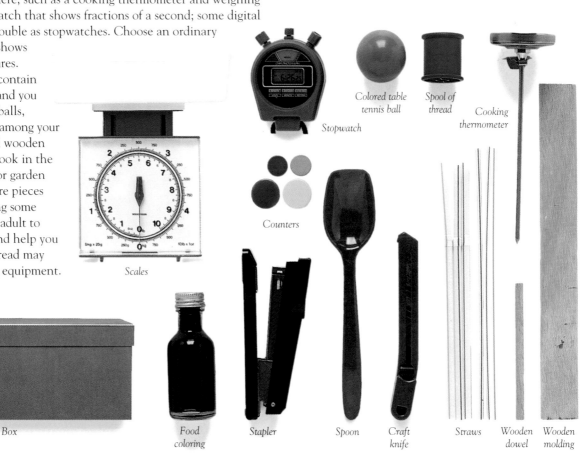

Stopwatch

Colored table tennis ball

Spool of thread

Cooking thermometer

Counters

Scales

Box

Food coloring

Stapler

Spoon

Craft knife

Straws

Wooden dowel

Wooden molding

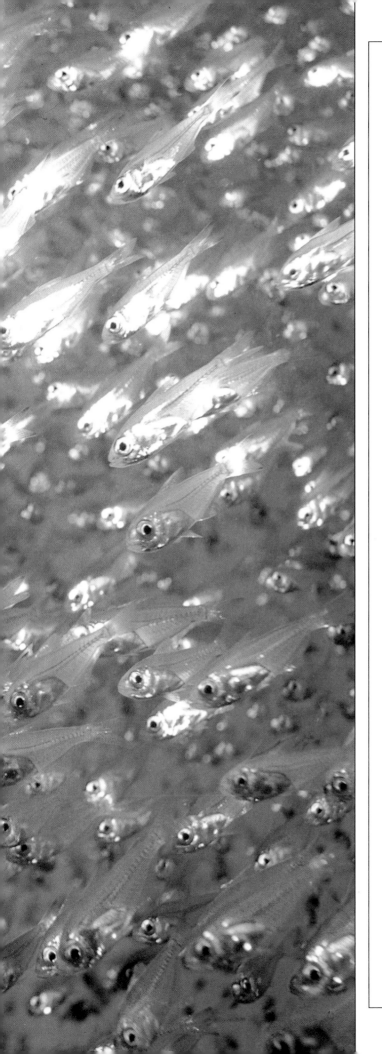

NUMBERS

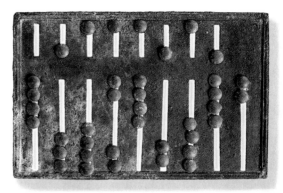

Masses and masses
The art of calculation, involving tools such as this Roman abacus (above), has developed over thousands of years. Using sophisticated computing, mathematicians can now work with large numbers like the number of hatchetfish in this shoal (left).

THE CONCEPT OF NUMBERS has been known to humans for millennia. Numbers were first used simply as a way of counting, to record amounts, but over the centuries mathematicians have found ways to manipulate them to gain new information. They have gone further, and invented symbols and words to define numbers and indicate how to use them. This language of mathematics has been a crucial tool in changing the world. It is now vital in many aspects of life, such as science, technology, and economics, as well as in music, philosophy, and entertainment.

THE FOUR SKILLS

JUST AS LETTERS ARE THE BASIS of writing, numbers are the main tools in mathematics. Arithmetic — working with numbers — has existed for more than 5,000 years. It now forms part of complex activities, from record keeping to the storage of electronic data. All calculations still rely on addition, subtraction, multiplication, and division.

Most people first come across numbers when they learn to count. The Western counting system is based on the number 10. It probably evolved because humans used their 10 fingers and 10 toes for counting. There are separate symbols, called digits, for the numbers 0 to 9, and larger numbers are shown as two or more digits written side by side. The number 27, for example, represents 2 tens and 7 units. This counting method is called a "base 10" or "decimal" system (p. 46). We do not know who first counted in base 10. However, historians who studied papyrus documents left by the ancient Egyptians have found that these people used a base 10 counting system nearly 5,000 years ago.

Placing digits

Although the ancient Egyptians counted in base 10, their number system differed from ours in many respects. For example, they did not form numbers from strings of digits, as we do, and therefore could not use a single digit, such as 5, in compound numbers, such as 50 and 500.

The earliest people known to write numbers by using different place values for digits were the Babylonians. Keen astronomers, they developed mathematics to help them predict the paths of the Sun, stars, and planets. Their way of writing numbers enabled them to devise far more complex techniques than the Egyptians. Clay tablets dating from about 1750 B.C. show that the Babylonians could calculate square roots and early forms of logarithms, and solve quadratic and cubic equations (p. 72).

Numerals

The Greeks grouped numbers in 10s, with symbols for 100 and other powers of 10 (p. 40). They came to adopt just the first letter of each symbol's name to stand for the number itself. For example, π (p, for *pente*) stood for 5, and Δ (d, for *deka*) represented 10. The Romans devised an even simpler system, based on the letters I, V, X, L, and C. (D, standing for 500, and M, for 1,000, were added later.) Their system lasted in Europe for nearly 2,000 years. Modern Western numerals evolved in India in the 7th century A.D. This number system, brought into Europe by

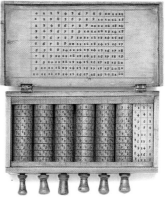

Napier's bones
These were developed as a calculating aid. This set is in the Old Royal Observatory, Greenwich, England.

Fibonacci (p. 30) in the 13th century, was far superior to any that had been developed until then.

None of the ancient peoples used zero (0). In the Babylonian counting system, for example, it is impossible to distinguish between numbers such as 82 and 802 because 0 was not used. Considering this, it is hard to imagine how the Babylonians developed mathematics as far as they did. The symbol "0" did not form part of any counting system until the 5th century A.D., when the egg-shaped 0 was first used by scholars in India. Three centuries later, the Arabs adopted 0 and took it to Europe.

Forming sums

The word "arithmetic" usually suggests the four skills of addition, multiplication, subtraction, and division (p. 16). Symbols for these processes are vital in mathematics. They have existed for thousands of years, as can be seen in the Rhind Papyrus (p. 14), a mathematical text from ancient Egypt. Its writer described the document as "directions for knowing all dark things," but it was essentially a mathematics tutor. The papyrus used special symbols for the four skills: a pair of legs walking forward, for example, showed

The Chinese abacus
The abacus has 13 columns of beads. Each has five beads to show 1s and two beads to show 5s. This calculating aid (p. 19) is still used in some parts of the world in place of the latest electronic calculators.

Mathematical symbols
Early symbols arose from trade as well as mathematics. The ancient signs here represent (left to right): +, −; ÷; and ×.

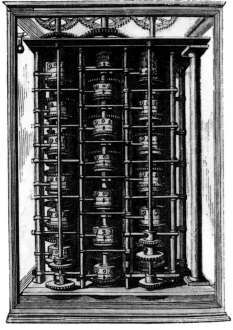

The difference machine
This completed section of Charles Babbage's "difference engine" shows cogs that would have been turned by steam to perform calculations.

addition; a pair of legs walking backward, or a flight of arrows, represented subtraction.

The earliest printed works in Europe showed the four processes with abbreviations of words, such as the Latin *et* ("and") which stood for addition. In 1557, an English mathematician, Robert Recorde (c. 1510–58), introduced the = sign for "is equal to;" he used two parallel lines because, in his words, "noe .2. thynges can be moare equalle." Other symbols for various operations were devised during the next few hundred years, but most of them could mean different things in the hands of different users. Today, with one or two exceptions (p. 17), symbols for operations have been standardized for worldwide use.

James Clark Maxwell
A Scottish physicist (p. 29), Maxwell devised a complex set of equations to describe how light, heat, and forms of electromagnetic radiation travel.

Techniques

In daily life, people often use mental arithmetic. For instance, when you buy something you have to work out mentally how much change you should receive from the money that you hand over. In general, the larger the numbers in a calculation, the harder it is to do mental arithmetic.

Many techniques and instruments have been used to make arithmetic easier. The abacus (p. 19) is one of the first devices; the earliest, from China, is more than 2,000 years old. There are many types around the world, all consisting of beads in rows or columns that signify digits in a number. Abaci are efficient tools for simple calculations, but not for multiplying and dividing large numbers. In 1614, John Napier invented logarithms (p. 22), which use the concept of powers (p. 40). Napier knew that it is possible to multiply numbers by adding their powers together. For example, $3^4 \times 3^2 = 3^{4+2}$ or 3^6. Logarithms, which are tables of powers of a given number, formed the basis for these calculations. He also invented "Napier's bones" (p. 23), a set of rods that were marked with numbers and worked on the same principle as his logarithms. A later version was the slide rule, a mechanical calculator that fell into disuse only when cheap pocket calculators became widely available.

Arithmetic machines

Blaise Pascal (p. 79) invented the first mechanical calculator, which was a type of adding machine, in 1642. But logarithm tables were still calculated by hand. In 1823 a British engineer, Charles Babbage (1792–1871), grew concerned about the number of errors hand calculation produced. He built a huge machine, which he called a "difference engine," to calculate logarithms and other tables automatically. Although the

engine was steam-powered, in many ways it resembled a modern electronic computer. It could be programmed to carry out a particular task, and would show the results on a legible print-out. Sadly, after 10 years' work, Babbage ran out of money. He had to stop work on the project before he had finished building his amazing machine.

The electronic computer, whose evolution was made possible by advances in physics, has helped scientists considerably to develop technology since the mid-20th century. It is very good at arithmetic. A computer can hold massive programs that enable it to perform many different tasks. Computers now feature in almost all areas of life. Many types are cheap and portable. By utilizing them, we rely on numbers and arithmetic more than ever before.

Using the pulley principle
Cranes lift and move heavy objects by making use of the pulley principle (p. 29). Engineers use mathematics to calculate the maximum load a crane can take.

Magnification
Scientists magnify very small or distant objects in order to see them more clearly. The magnifying glass (above) is one of the most basic instruments employed for this purpose, and is very easy to use.

Ways of writing numbers

EARLY CONCEPTS OF NUMBERS can be seen in artifacts 30,000 years old, but only in the past 6,000 years have humans recorded their calculations in writing. From 3000 B.C. the Egyptians were using fractions, and from 2000 B.C. the Babylonians used place values (p. 46) for the digits in numbers. In the Classical age in Greece (c. 600 B.C.–A.D. 200), arithmetic was a highly regarded discipline. The work of the Greek mathematicians — Aristotle, Euclid, Plato, and Pythagoras — yielded vital information. A new system was developed which used letters to represent numbers, like the Roman numerals we see today. But it was not until around A.D. 600, in India, that the system of 10 symbols, together with place values of hundreds, tens, and units, became common.

Egyptian papyrus

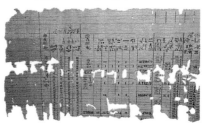

The Rhind Papyrus, from Egypt, rediscovered in 1858, was copied by a scribe named Ahmes around 1650 B.C. However, the material it contains dates from hundreds of years before that. It has 84 arithmetic and algebraic problems together with their solutions. One set of symbols is drawn like walking human legs — facing forward, it means "add," and facing backward, it means "subtract."

EXPERIMENT
Simple number puzzle

In A.D. 830, the Arab scholar al-Khwarizmi (p. 68) wrote a major work on Hindu Indian numerals. So important was this work in medieval Europe that Western numerals are commonly, and wrongly, still known as "Arabic." The tenth symbol, zero (pp. 30–31), was represented by a round goose egg. The puzzle below includes all 10 numerals. You can make the puzzle to help younger children learn their numbers.

You Will Need
● *ruler* ● *pencil*
● *pen* ● *scissors*
● *glue* ● *paper*
● *poster board ½ in (1 cm) larger all around than the paper*

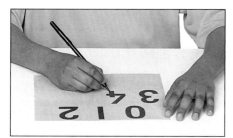

1 DIVIDE THE paper into three equal sections horizontally and draw in the numerals 0–9. You could create these numerals on a computer. Type in the numbers and then enlarge the font you use to at least 150 points in size.

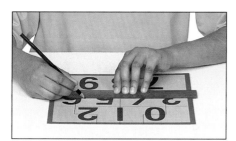

2 APPLY glue to the back of the numbered paper. Lay the paper centrally on the poster board. Use a ruler to draw random shapes over the numbers.

3 CUT along the pencil lines with scissors to make the puzzle pieces. Now muddle them up and give them to a child to reassemble.

Doing the puzzle
The puzzle is not as easy as it looks, although the frame gives a useful guide to help in assembling it. The puzzle works best if each numeral is cut into two pieces only.

EXPERIMENT
Chinese puzzle

Mathematicians in ancient China loved puzzles. The one below uses Chinese numerals. First, make the puzzle, cut it up, and then try to put it together again. See if you can work out what each symbol stands for and how larger numbers are built up from them. (To help you, the number 10 looks rather like a T-shaped cross.) Once you understand how the pattern is formed, you can "read" the numbers and easily assemble the puzzle.

YOU WILL NEED
- water ● glue ● *craft knife* ● *paintbrush*
- *paper* ● *foamcore* *½ in (1 cm) larger all around than the paper*
- *paints* ● *cutting mat*

 Adult help is advised for this experiment

Use bold brush strokes to paint the numbers

The assembled puzzle

Chinese lettering
Photocopy the Chinese numerals shown on the right, and use the photocopy to trace the outlines onto a piece of paper. Paint in the Chinese characters. When the paint has *dried, mark the lines to make the puzzle. Glue the paper onto the foamcore. Ask an adult to cut along the lines with a craft knife to create the puzzle pieces.*

Unlucky thirteen

The superstitious fear of the number 13 is called triskaidekaphobia. In Christianity, 13 is linked with the Last Supper of Jesus and his 12 disciples, shown in this copy of a picture by Leonardo da Vinci. The thirteenth person, Judas, betrayed Jesus. Some people refuse to take any risks on the 13th day of the month. However, no proof exists that 13 is unlucky.

Evolving number systems

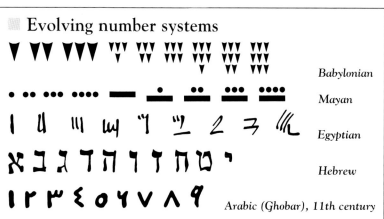

The numbers we find so familiar today have evolved over thousands of years, through conquest, technology, and writing. Babylonian cuneiform is one of the oldest written number systems known. The Mayan system was developed independently in the Americas, and is thought to date from before 500 B.C. Modern Western numbers have their origins in both Hindu script and Arabic numerals, which came to Europe as a result of the writings of al-Khwarizmi (above left).

Using numbers

FOR THOUSANDS OF YEARS mathematicians have worked to discover the hidden properties of numbers. To help with counting they developed arithmetic — from the Greek word *arithmetike*, meaning the art of numbers. Arithmetic uses four simple operations: addition, subtraction, multiplication, and division. The basic operation, addition, is a way of putting numbers together to make bigger numbers. It is repeated counting forward, and subtraction is repeated counting backward. Each of the operations is directly linked to the others. Subtraction is the opposite, or inverse, of addition; multiplication is a quick way of doing repeated addition; division is the inverse of multiplication. There is a particular shorthand symbol for each. These simple building blocks are one of the bases of our understanding of mathematics.

Inscribed stone
Cuneiform script was developed in Mesopotamia (now modern Iraq) during the 4th millennium B.C. It used pictures and marks to record information and numbers. Records were pressed into soft clay tablets with a stamp, then the tablets were baked in ovens or in the hot sun to save the information. Some of these tablets date back to 3000 B.C. and show the Babylonians' use of base 60 (p. 46). This tablet is dated 2900 B.C. and shows information about crop yields. It is written in vertical columns reading from right to left.

123 Trick

How do you measure four units of water when you have only three- and five-unit jugs? To find out, you will need a jug that takes three units and one that takes five units. Have a large bowl nearby. The unit can be a cup or glassful.
1. Fill up the three-unit jug and pour the water into the five-unit jug.
2. Fill up the three-unit jug again with more water and pour the water into the five-unit jug until it reaches the top. You will be left with one unit in the three-unit jug.
3. Empty the five-unit jug into the bowl and pour the rest of the water from the three-unit jug into the five-unit jug.
4. Fill the three-unit jug again and add it to the water in the five-unit jug. This will make four units. Now check your result.

Using scales

The two most basic skills in arithmetic are addition and subtraction. What, for example, must be added to 3 to give 10? The answer is 7, and 7 is called the "complement" of 3. In other words, 7 is equal to $10 - 3$. If you are very new to adding and subtracting, you can use scales to help you to get a grasp of simple arithmetic.

Add one block to the scales

1 WHEN one block is placed on the scales, the total weight is one unit, and the scales read 1 to show this amount.

Add another two blocks to the scales

2 BY putting another two blocks on the scales, you bring the total amount to three units. You have added 1 and 2 together.

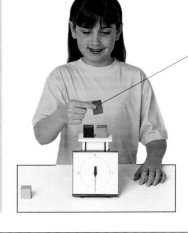

Now remove one block from the scales

3 WHEN one block is taken off the scales, the total is back to two units because one has been taken away, or subtracted.

❓ Gray matter

Find the odd one out in each of these sets of sums using a calculator. (Answers on p. 186.)

1 (a) 93+145+12 (b) 175−8+83
(c) 153+124−32

2 (a) 215×12 (b) 50×53 (c) 43×60

3 (a) 4032÷63 (b) 320÷5 (c) 804÷12

Ancient symbols

Just as the symbols for numbers have changed over the centuries (p. 14), so have those for mathematical operations. Number symbols have become virtually universal, particularly with the use of standard computer keyboards worldwide. Symbols for operations sometimes do differ: for example, some people use × as the sign for multiplying, while others use a point (.).

Addition
This Renaissance symbol is based on the Italian word for plus, "piu." Our + sign is a version of the Latin word for and – "et."

Subtraction
This is the Greek sign for minus. Our minus sign (−) probably derives from symbols used by traders to show differences in weights.

Multiplication
Leibniz (p. 47) used this sign for multiplying because the × was similar to a letter in algebra (pp. 70–71).

Division
This sign was used in 18th-century France. Ours may come from a fraction drawn with dots instead of numbers.

❓ Gray matter

The signs in these sums have been left out. Can you guess what they are? Check your answers on a calculator. (Answers also on p. 186.)

35	64 = 99	22	41 = 63
60	15 = 4	19	3 = 57
75	60 = 15	121	11 = 11
999	333 = 666	7	63 = 441

Game of NIM

In the ancient game of NIM, 21 coins or counters are laid out, and each of two players can pick up one, two, or three counters in turn. The player who picks up the last counter loses the game. You can win by using simple mathematics — in this case addition and subtraction. Make sure that at the end of each round a total of four counters has been picked up. You will always have control over the last turn, when there will be four counters, or fewer, left.

YOU WILL NEED
● *counters*

1 LAY THE COUNTERS out and tell your friend the rules (but not how to win). Ask her to go first. When she has taken her counters, you then take enough counters to reduce the total number by four. In this game, when the girl takes three counters, the boy takes one.

The second to last player has no chance of winning

2 CONTINUE WITH the game, making sure that a total of four has been picked up at the end of each round. You can play with as many counters as you like, as long as the number is always a multiple of four. The loser will be the person who picks up the last counter.

A history of calculation

FOR THOUSANDS OF YEARS people have used complex calculating methods. The Egyptian system, dating from around 2000 B.C., was clumsy, with multiplication and division done by repeated addition and subtraction. At the same time the Babylonians devised a superior system, using tables for multiplication and for square and cubed numbers. These systems lasted for centuries, until 600 to 400 B.C., during which the Greek mathematicians Pythagoras (p. 124), Thales, and Plato (p. 152) made a distinction between arithmetic (the theory of numbers) and logistic (the technique of calculation, seen as vital for traders and soldiers). At about the same time, the Chinese adopted an abacus using a decimal system. The Hindus developed ways for performing multiplication and long division, and it was their system that was adopted by the Arab world around A.D. 800. Since then calculation methods have been further refined, mainly for use in commerce and administration.

■ DISCOVERY ■
Archimedes

Archimedes (c. 287–212 B.C.) is considered to be the greatest mathematician of antiquity. He lived in Syracuse (Sicily, Italy) after studying for a short while in Alexandria, Egypt. He was a highly practical man, and many mathematical inventions are attributed to him, including Archimedes' screw (p. 148). Among other achievements, he developed the laws that govern floating bodies (p. 62), a method for achieving fantastically large calculations, an accurate estimate of pi (p. 134), and numerous geometrical formulas. His formula for the volume of a sphere (p. 104) was inscribed on his tombstone after he was killed by a Roman soldier.

░ Using fingers for calculation

Fingers were first used as sign language to express simple numbers (p. 92). You can use them to help you with the 9-times table. Hold both hands in front of you, and number the fingers, in your mind, from 1 (thumb on your left hand) to 10 (thumb on your right hand). To calculate a single digit, say 7, multiplied by 9, bend down the finger for that digit, as shown below. The fingers to the left of the bent finger show the tens, and those to the right show the units. So $7 \times 9 = 60 + 3 = 63$.

60   **3**

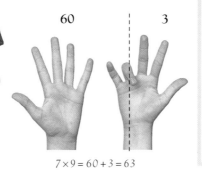

$7 \times 9 = 60 + 3 = 63$

123 Trick

1. Give a friend a calculator and tell her to keep it out of your sight. Without letting her see, write 9 on a piece of paper and put it face down on the table. Ask her to key in the last two digits of her telephone number.
2. Now instruct her to multiply the number by 2 and press =.
3. Add the number of her house or flat to this total and press =.
4. Subtract her age, and press =.
5. Multiply the total by 18; press =.
6. Ask your friend to add together all the digits in the answer. (For example, for 756, $7 + 5 + 6 = 18$.) Get her to cancel the previous number and key in this new addition. She should continue to add the digits of each new answer until she has a single digit. Now reveal your answer. Is it the same as her answer?

The abacus

The word "abacus" is from a Semitic word *ibq* meaning "to wipe the dust." The sand surface used for writing evolved into a board marked with lines, on which counters were used to indicate numbers. In China and Japan, where it is still used, the abacus is a frame with beads strung on parallel wires and a bar across the middle. Each of the two beads on the upper part of the wires represents five units, and each of the five beads in the lower section is one unit. Each wire represents units, tens, hundreds, and so on, with the lower values to the right. In some cases, the frame has a little dot representing a decimal point and marks like the commas in 1,000,000. To show a number, the appropriate beads are moved to the separating bar. Chinese users say that using an abacus can be faster than working with a calculator.

Using a Chinese abacus

The abacus is laid flat on a table, with the groups of five beads toward the user. When forming numbers, the user flicks the beads toward the separating bar with thumb and forefinger.

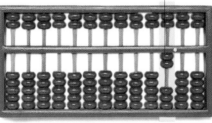

3 units | One 5 | Decimal point

1 To WORK OUT the sum of 8 + 4, the abacus first shows the decimal 8.00.

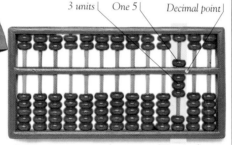

Subtract a 5 bead and 1 unit

2 To ADD 4, first subtract 6 (4+6=10), from the units column. This leaves only 2 in the units column.

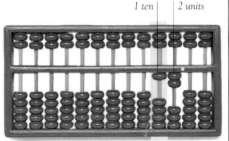

1 ten | 2 units

3 ADD 10 TO give the answer. To do this, move a bead up in the tens column. The abacus gives the answer when you read off the beads: 8 + 4 = 12.

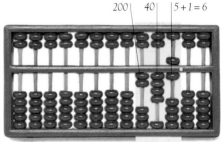

200 | 40 | 5+1=6

1 To WORK OUT the answer to the sum 246 + 372, the abacus first shows the decimal number 246.00.

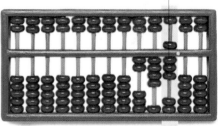

2 added in the units column

2 ADD 2 ON the units column by flicking up 2 beads to make 248.

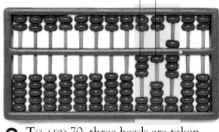

Add 100 | Take off 30

3 To ADD 70, three beads are taken from the 10s column and one is added to the 100s to make 318.

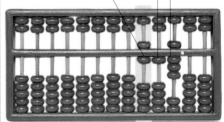

500+100=600 | 10 | 5 and 3 units=8

4 To ADD 300, 1 bead has been moved down to the separating bar on the 100s wire and 2 beads have been subtracted beneath to give the answer.

The calculator

The first mechanical calculator was devised by Blaise Pascal (p. 48) in 1642. Mechanical calculators were used for laborious arithmetical tasks until the introduction of the electronic calculator in the 1960s. Modern calculators are inexpensive and they can do both simple and scientific calculations.

Controlling numbers

MATHEMATICIANS HAVE ALWAYS used shortcuts to help them complete or check complicated sums. By the 12th century A.D. the Hindus in India had developed a system close to the one we use today, based on a knowledge of the patterns in numbers. Today calculators and computers do sums in an instant. But even complex calculations can be broken down and worked out on paper if you understand the relationships between operations. For example, multiplication is simply a quick way of adding numbers together. It is also linked to division, where one number is subtracted from another one several times. By knowing the tricks of calculating you will begin to be able to control numbers.

■ Looking up tables

Tables of numbers, like nautical almanacs, have been used for more than 4,000 years as a reference guide when completing complex calculations. Some, such as those for trigonometric values and logarithms, are now worked out by calculators, but it is easiest to see the relationships between the numbers by studying the tables themselves.

Nautical almanac
This almanac was first published in 1766 for navigators. The tables show distances between stars and the Moon at three-hour intervals, helping sailors to find their position when at sea.

INVESTIGATING NUMBERS
Halving and doubling

The 2-times table can help you check difficult multiplication; you halve and double some simple numbers. The sum here is 33×86.

1 Make two columns and write down the numbers opposite each other.

33	86

2 Divide the numbers in the first column by 2, ignoring any remainders until you get to the number 1.

| 16 |
| 8 |
| 4 |
| 2 |
| 1 |

3 Now multiply the numbers in the second column by 2 until it is level with the first column.

33	86
16	172
8	344
4	688
2	1376
1	2752

4 Cancel every number in the second column (on the right) that is opposite an even number in the first one.

33	86
16	~~172~~
8	~~344~~
4	~~688~~
2	~~1376~~
1	2752

5 Add the numbers that are left over in the second column to give the answer to the calculation.

$$2752 + 86 = 2838$$

🧩 Puzzle
Try dividing the number 1274953680 (a number made up of all the numerals) by any number up to 16. What do you find? What do you notice about this number? (Answers on p. 186.)

INVESTIGATING NUMBERS
Casting out nines

This is a way of using the number 9 to check if you have worked out a long multiplication calculation correctly. Look at the sum $8326 \times 6439 = 53611114$.

1 Add the digits of each number together.

$$8 + 3 + 2 + 6 = 19$$
$$6 + 4 + 3 + 9 = 22$$
$$5 + 3 + 6 + 1 +$$
$$1 + 1 + 1 + 4 = 22$$

2 Cast out 9's in the three numbers given by dividing by 9 and noting down the numbers left over.

$$19 \div 9 = 2 \ (1 \text{ left over})$$
$$22 \div 9 = 2 \ (4 \text{ left over})$$
$$22 \div 9 = 2 \ (4 \text{ left over})$$

3 Multiply the left-over numbers of the first two sums. The result should equal the third left-over number.

$$1 \times 4 = 4$$

EXPERIMENT
Book of permutations

A small number of items can be grouped to give a large number of permutations. By making a book of eight pictures, each divided into three sections, you can see how the permutations increase with each additional page. Templates are provided for you to enlarge and copy.

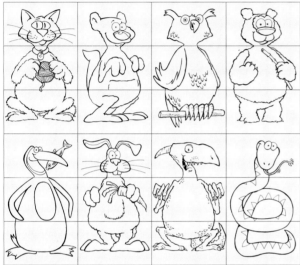

Template for pages of book

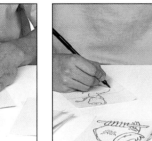

1 CUT OUT eight equal-sized pieces of paper. Rule pencil lines across the paper to divide each sheet into three horizontal sections. From the poster board, make two covers the same size as the pages.

2 COPY one animal on each page, leaving space at the left for the book's spine. Use the ruled lines as guides so that each shape fits within the sections on the page. They will line up when the pages are cut.

3 COLOR all the animals, then put the eight pages between the two pieces of poster board. Line up the edges, and staple through all thicknesses to make the book.

INVESTIGATING NUMBERS
Finding out about dividing

There are many clever ways of finding out whether a number is divisible by another just by looking at the digits in it.

DIVISIBLE BY 10
The number ends in a 0, which can be crossed off.

$$1200 \div 10 = 120$$

DIVISIBLE BY 4
The last two digits in the number are 00, or are divisible by 4.

$$1644 \div 4 = 411$$
$$(44 \div 4 = 11)$$

DIVISIBLE BY 3
The digits in the number add up to a multiple of 3.

$$72 \div 3 = 24 \quad 192 \div 3 = 64$$
$$(7 + 2 = 9) \quad (1 + 9 + 2 = 12)$$

YOU WILL NEED
- *calculator* ● *ruler* ● *pencil*
- *pen* ● *scissors* ● *stapler*
- *poster board* ● *paper*

4 USE THE SCISSORS to cut along the horizontal lines through all thicknesses, except the cover. Open the pages at random and see what you find. Can you work out mathematically how many different animals are made? (Answer on p. 186.)

Calculating aids

As all knowledge has grown over the centuries, the ability to express and use new information mathematically has become ever more important. Many devices have been used to help people perform complex calculations. Logarithms were invented in 1614. A logarithm is the power (p. 40) of a number in a particular base — usually base 10. For example, $1000 = 10^3$, so the log of 1000 is 3. Using logarithmic tables, multiplication could be done simply by adding logarithms and referring to tables to find the answer. From the abacus (p. 19) to Napier's bones and electronic calculators, computational aids have allowed numbers to be processed with increasing speed and accuracy.

▓ What are these?

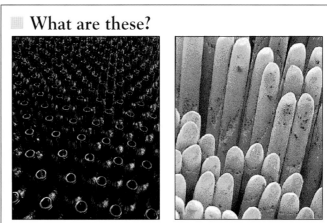

These two photographs are of everyday objects that have been photographed at different magnifications. Can you guess what they are? (Answers on p. 186.) It is important to know the magnification of an object in order to identify it. For example, part of the Moon photographed through a telescope could look, to the untrained eye, like a piece of rock or a lump of plaster under a powerful scanning electron microscope, if the magnification was not known.

EXPERIMENT

Calculating magnification

When you use a magnifying glass to look at an object, the image you see through the glass is bigger than the object itself. To investigate if one magnifying glass is more powerful than another, you need to use mathematics — measuring and dividing. The magnification of a glass is found by dividing the length of the magnified image by the length of the actual life-size object.

You Will Need
- *calculator*
- *notepad*
- *pen*
- *small object*
- *magnifying glass*
- *graph paper*

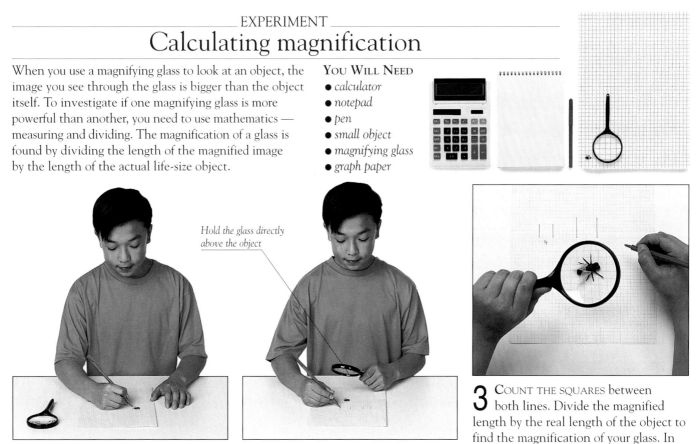

Hold the glass directly above the object

1 Place the object (here it is a plastic insect) on the graph paper, and mark the paper at both ends of its body to determine the life-size length.

2 Move the object down and to the right and look at it through the glass. Mark on the paper, level with the first marks, the apparent length of the object.

3 Count the squares between both lines. Divide the magnified length by the real length of the object to find the magnification of your glass. In this experiment the insect (a fly) covered four squares on the paper at life-size, and six squares when magnified; so this glass magnifies objects 1½ times.

EXPERIMENT
Making Napier's bones

Napier's bones, a set of rods with a series of numbers inscribed on them, were devised as a simple means for calculating. Set in a container, side by side, they could be used for doing complex multiplications and divisions by means of simple addition and subtraction. Napier's bones were used for more than 300 years, until the advent of mechanical calculators and then electronic calculators.

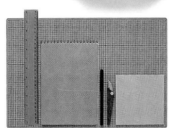

1 MEASURE A 5½-in (14-cm) square on a piece of foamcore. Ask an adult to cut it out for you. Now measure and rule off nine evenly spaced columns across the square and nine rows down.

Adult help is advised for this experiment

YOU WILL NEED
- cutting mat • ruler
- notepad • pen
- craft knife • storage box • foamcore

2 WRITE THE NUMBERS 1 through 9 at the tops of the nine columns. Draw a diagonal line from the top right corner of each second row through opposite corners of the other boxes as shown.

3 WRITE THE NUMBERS 2 through 9 down the left-hand side. Multiply each number at the top by each number on this side; put the 10's in the top corner of a box and the units in the bottom corner.

4 ASK AN ADULT to cut along the vertical lines to make the nine "bones." Store them in a box for safety until you are ready to do a complex sum.

5 REMOVE BONES with the numerals you wish to multiply. To find out the answer to 1572×3, take out the bones headed with the numerals 1, 5, 7, and 2.

■ DISCOVERY ■
John Napier

Napier (1550–1617) was a Scottish landowner who studied mathematics as a hobby. He had a special interest in trigonometry and computation. Napier invented logarithms to simplify multiplication and division and published *A Description of the Marvellous Rule of Logarithms*, which contained the first logarithmic tables, in 1614. The tables had taken 20 years to complete. He created his system of numbered rods, Napier's bones, soon afterward. These provided a new way of doing calculations simply. He also made the use of the decimal point (pp. 34–35) generally popular in Europe.

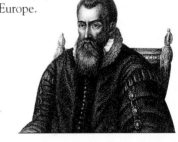

The units are in the bottom corner

The 10's are in the top left corner

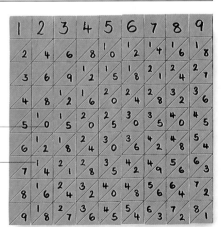

The finished bones

1572×3
Lay the bones so that the tops show 1572. To multiply this number by 3, add up each pair of numbers along the diagonals of the third line. Starting at the right, write the numbers: 6=6 (the last digit in the answer); 0+1=1; and so on. (Answer on p. 186.)

Using your brain

NOWADAYS, WITH THE COMMON USE of calculators and computers, some people think that skill in mental arithmetic is no longer necessary. However, this skill is still important because it encourages a genuine understanding of numbers. Good mathematicians always look for the simplest methods and rules and attempt to reduce complicated calculations to a series of basic sums. These methods need practice, but eventually will become invaluable aids that you use automatically.

Mathematical genius

Shakuntala Devi is an Indian woman of immense arithmetical ability who demonstrates her amazing skills in public. It took just a second for her to find that the cube root of 332,812,557 is 693. In the 1970s she calculated the 23rd root of a 201-digit number. The problem had been devised by Texan students. A computer took one minute and 13,000 instructions to check her answer and prove her right.

INVESTIGATING NUMBERS
Adding up

To add numbers together in your head using mental arithmetic, learn to round off at least one of the numbers to the nearest 10.

1 In the sum 56 + 33, add 4 to 56 to round it off to 60.

$$56 + 4 = 60$$

2 Now subtract the 4 from the other digit in the sum (33) to balance the sum.

$$33 - 4 = 29$$

3 Now do the simplified sum by mental arithmetic.

$$60 + 29 = 89$$
$$(56 + 33 = 89)$$

INVESTIGATING NUMBERS
Working with 5

Dividing or multiplying by the number 5 can be made easier and quicker if you recognize that the number 5 has a relationship with the number 10 (see below).

1 To work out the answer to 245 ÷ 5, first multiply by 2.

$$245 \times 2 = 490$$

2 Then divide the answer by 10.

$$490 \div 10 = 49$$
$$(245 \div 5 = 49)$$

INVESTIGATING NUMBERS
Simpler subtraction

A difficult subtraction can be made easier by rounding off one of the numbers, up or down, to the nearest 10 or 100.

ROUNDING OFF
For 67 − 18, round off the 18 to 20. Add the same number, 2, to 67.

$$18 + 2 = 20$$
$$67 + 2 = 69$$
$$69 - 20 = 49$$
$$(67 - 18 = 49)$$

CHECK YOUR ANSWER
Add your answer to the number first subtracted.

$$67 - 18 = 49$$
$$49 + 18 = 67$$

THREE-DIGIT NUMBERS
For 651 − 177, round off 177 to the nearest hundred. Add the same number, 23, to 65.

$$177 + 23 = 200$$
$$651 + 23 = 674$$
$$674 - 200 = 474$$
$$(651 - 177 = 474)$$

INVESTIGATING NUMBERS
Multiples of 10

The number 10 is probably the easiest number to work with in mental arithmetic. Division and multiplication by 10, or any multiples of 10, involve only a shift in place value for the digits rather than any change in the digits themselves.

DIVIDE BY 10
Shift digits one place right or move decimal point (p. 35) one place left.

$$3785 \div 10 = 378.5$$
$$235 \div 10 = 23.5$$

DIVIDE BY 100
Move the decimal point two places to the left.

$$4568 \div 100 = 45.68$$
$$5600 \div 100 = 56$$

MULTIPLES OF 10
Shift digits the required number of places to the left and put on the correct number of zeros.

$$23 \times 10 = 230$$
$$245 \times 100 = 24,500$$
$$987 \times 1000 = 987,000$$

Backward and forward

Numerals or words that are the same backward as forward — such as 1991 or the word "madam" — are known as palindromes. This trick, using the idea of palindromes, will help you practice arithmetic.

YOU WILL NEED
● *notepad* ● *pens*

1 ASK A FRIEND to look away while you write the number 1089 on a piece of paper. Fold the piece of paper and leave it where you cannot touch it but where your friend can still see it.

2 ASK YOUR FRIEND to write a non-palindromic three-digit number. Reverse the number and subtract the smaller number from the larger. Then reverse that answer and add it to the previous one.

3 THE answer will always be 1089, the number that you have written on the paper. You can unfold your answer to show to your friend and show him how clever you are.

Magic with 11

There are several ways of calculating mentally with the number 11. Here is one way to multiply by 11 and one to see if a number is divisible by 11.

TO MULTIPLY A NUMBER BY 11
First multiply the number by 10, then add the original number to it.

$$865 \times 11$$
$$865 \times 10 = 8650$$
$$8650 + 865 = 9515$$

TO FIND NUMBERS DIVISIBLE BY 11

1 To find out if any number is divisible by 11, start with the digit on the left, subtract the next digit from it, add the next digit, subtract the next, and so on.

$$53746$$
$$5 - 3 + 7 - 4 + 6 = 11$$

2 If the answer is 0 or 11, then the original number is divisible by 11.

$$53745 \div 11 = 4886$$

🧩 Puzzle
Without hesitation, can you write down the result of eleven thousand plus eleven hundred plus eleven quickly? (Answer on p. 186.)

Reducing to 9

In any number divisible by 9, the digits in the number add up to a number divisible by 9. For example, $4 \times 9 = 36$ and $3 + 6 = 9$.

1 To see if a number (e.g. 201,915) is divisible by 9, add the digits together.

$$2 + 0 + 1 + 9 + 1 + 5 = 18$$

2 Keep adding digits together until you arrive at a single figure. If this figure is 9, then the number is divisible by 9.

$$1 + 8 = 9$$

Multiplying with square numbers

This is a quick way to multiply numbers that differ by 2. It works best if the number in between can easily be squared in your head (pp. 40–41).

1 To multiply 29 by 31, for example, first mentally square the number in between the two: in this case, 30.

$$29 \times 31$$
$$30 \times 30 = 900$$

2 When you have squared the number, subtract 1 from the result to find the answer to the calculation.

$$900 - 1 = 899$$

Having fun

YOU CAN learn about the special quirks, characteristics, and peculiarities of numbers by having fun with them. Tricks, puzzles, and number patterns have tested the intellect of mathematicians for thousands of years (pp. 172–173). These brain teasers have helped to advance our understanding in many areas, from logic to the study of shapes found in nature.

123 Trick

1. Ask a friend to choose any number (no more than five digits), and to write it down where you cannot see it.
2. Ask your friend to multiply this number by 9 — mentally or on a calculator — then add his or her age to this total. Then ask your friend to tell you the final total. You write this total down. Tell your friend that you will now be able to say how old he or she is.
3. Do the following calculation, either in your head or where your friend cannot see. (With practice you will be able to do this quickly, which will look more impressive.) Add all the digits of the number together, until you have a two-digit number.
4. Now add the two digits together to get a one-digit number. If your friend appears to be under 10 years old, this single digit is his or her age. If your friend looks older than 10, increase the total by adding 9 each time until you reach a likely age.

Puzzle

1. Copy the 3×3 "magic" square and write in the numbers shown inside.
2. Using the numbers from 4 to 12, complete the square so that the totals of each row, column, and diagonal are the same. Write the totals just outside the square, at the end of each line. (Answer on p. 186.)

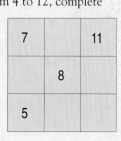

7		11
	8	
5		

? Gray matter

This is a way of making your own personal birthday square.

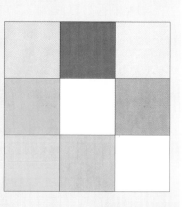

1. Make a 3×3 square based on your own date of birth or that of a friend. The colored template gives you a guide to the squares to be filled in.
2. Select a date of birth, and write it outside the empty square. Sample date: 26/3/84.
3. Write the two-digit year number (84) in the red square.
4. Add the year number and the day number together and put the total into the gray square.
5. Add the day number to this total, and write the new total in the purple square. The numbers in these first three positions form the basis for the rest of the square.
6. Subtract the month number from the number in the red square. Write the total in the white square. Subtract the month number from this new number, and put the answer in the orange square.
7. Subtract the month number from the number in the gray square, and write the answer in the yellow square. Subtract the month number from this total, and put this new number in the blue square.
8. Subtract the month number from the number in the purple square, and write the answer in the green square. Subtract the month number again from this total, and write the new answer in the pink square. You now have a number in each square.
9. Add the numbers in each row, and the numbers in each column, and those on each diagonal of the square; the total for every line should be the same. This can be your "lucky number."

EXPERIMENT
Spiral board game

Number games include snakes and ladders. The spiral board game below, which is drawn on graph paper and built up like a magic square, allows you to make your own rules using numbers and moves of your choice.

The rules of this game are that players take turns and move a counter forward at each throw of the dice. The first player to land on 100 is the winner.

YOU WILL NEED
- colored pencils ● pens
- scissors ● glue stick ● 2 dice
- counters ● ruler
- foamcore ● graph paper

1 DRAW A GRID of 100 fairly big squares. Count down five rows and across five squares, and write 1. Put 2 in the square to the right, 3 below 2, and so on in a spiral.

🔢 Trick

1. Ask a friend to write two six-digit numbers, one below the other. Call the rows **a** and **b**.
2. For each digit in row **a**, find another digit (called the complement) that can be added to it to make up 9. These digits form row **c**.
3. Write in any six-digit number for row **d**.
4. Using row **d**, make the digits in row **e** the complements of 9 as before.
5. Now you can do the sum of these five rows (row **f**) quickly before your friend's eyes.
6. The sum of all the rows will be equal to the digits in row **b** with 2 subtracted and the digit 2 placed at the front.
   The pairs of complements total 1,999,998, which is 2,000,000 minus 2. Therefore, the answer will always be the random row (row **b**) plus 2 million less 2.

🔢 Trick

1. Write down any number up to 30 — say 25.
2. Multiply the second digit by 4, then add on the first. Use the new number to create another in the same way. Write them in a circle. You will return to the original number.

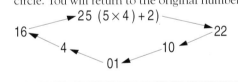

$$16 \leftarrow \quad \nearrow 25 \ (5 \times 4) + 2)$$
$$22$$
$$4 \qquad \qquad 10$$
$$01$$

🧩 Puzzle

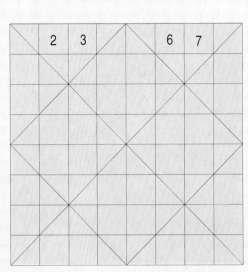

1. Copy the 8 × 8 square shown in the diagram. Draw diagonal lines from corner to corner, and between the central points of each side of the square, as shown.
2. Now fill the square with the numbers 1–64 as follows. Start entering numbers at the top left-hand corner, but skip any number that falls in a square with a line through it. Here, the first row has been done for you.
3. Begin again from the bottom right corner of the square. Fill in the numbers in order, starting with the number 1, inserting numbers only in the squares with lines through them.
4. Add the numbers on each row, column, and diagonal. Write the total just outside the big square, at the end of that line of numbers.
5. Add the numbers in each corner of the square to the four numbers in the center of the square. As you can see, the diagonal lines between the center points of the big square's sides form a smaller square. Add the lines of numbers on each side of this small square. Write the totals outside the big square at the end of each line. Now add the totals of the opposite sides of the smaller square. What do you notice about the answers? (Answer on p. 186.)

2 WITH A COLORED PENCIL, shade in a spiral path through the center of each square from 1 to 100. Shade the edges of the path in a contrasting color.

3 ON THE FOAMCORE, rule a square slightly larger than your grid. Cut out the square, then glue the grid on top, leaving a narrow frame around it.

4 TO PLAY the game with a friend, make your moves according to the numbers on the dice. The winner is the one who reaches 100 first. You can invent your own rules; for example, if you land on a multiple of 7, you have to miss a turn, or you could multiply the numbers on the dice by 2 to make the game shorter.

73	74	75	76	77	78	79	80	81	82
72	43	44	45	46	47	48	49	50	83
71	42	21	22	23	24	25	26	51	
70	41	20	7	8	9	10		52	
69	40	19	6	1	2	11	28		
68	39	18	5	4	3	12	29	54	
67	38	17	16	15	14	13	30	55	
66	37	36		34	33	32	31	56	89
65	64	63	62	61	60	59	58	57	90
100	99	98	97	96	95	94	93	92	91

Each player has one counter

Mathematics and science

MANY OF THE WORLD'S great mathematicians, including Archimedes (p. 18) and Newton (p. 71), were also scientists, using their reasoning skills and logic to cross the boundaries between physical science and mathematics. They used mathematics to express relationships between physical quantities in such a way that we are able to make predictions. You may notice that a pendulum swings faster the shorter it becomes (p. 109). Through mathematics, the relationship between the pendulum's length and the time taken for one swing can be clearly defined. As the study of mathematics has progressed, so too have science and technology. All scientific formulas are mathematical; the application of formulas, or mathematical modeling, is widely used for diverse functions, from long-range weather forecasting to money trading (p. 181). Without mathematics, they could not even be considered.

A mathematical bridge

The wooden bridge in Queens' College, Cambridge, England, is known as the mathematical bridge — possibly because of the mathematical concepts behind its design. The base is an arc of a circle (p. 136), and the load is carried on beams laid at a tangent (p. 185) to the arc. The pieces that are topped with the handrail are radii in the arc of the circle. The original bridge was built in 1750. The present bridge, made from teak, was built in 1904.

EXPERIMENT
Acceleration

Galileo (1564–1642) used this experiment to find a formula for acceleration (abbreviated to *a*). The formula showed that the distance (*s*) traveled by an accelerating object (which was stationary to start with) is proportional to the square of the time (*t*) elapsed (acceleration is equal to twice the distance divided by the time squared; $a=2s/t^2$). He used mathematics to investigate this physical principle, timing the rolling of a ball with his pulse. The experiment shows that acceleration remains constant for an object rolling down a slope. It is only the angle of the slope that affects the acceleration.

YOU WILL NEED
● *ruler* ● *notepad* ● *ball* ● *modeling clay*
● *pens* ● *stopwatch*
● *piece of molding*
at least 3 ft (1 m) long

1 MEASURE the piece of molding and, using a pen, mark a point on the molding every 4 in (10 cm), working from one end to the other.

2 USE a piece of modeling clay to raise the top end of the molding about 1 in (2.5 cm) and make an incline. The modeling clay will also hold the wood firm.

3 ASK A FRIEND to work the stopwatch and note the time it takes you to roll the ball down from the top. Time how long it takes to roll the first 12 in (30 cm) and write this figure down. Next, time the ball rolling 24 in (60 cm). To find the acceleration for each distance, multiply the distance by 2 and divide this figure by the time squared (p. 40). What happens to the acceleration over a number of distances? (Answer on p. 186.)

EXPERIMENT
Making a pulley

A load attached to one end of a rope can be lifted by a pull (called the effort) on the other end. Multiple pulleys give increased lifting power, which is known as the mechanical advantage. This is found by dividing the weight of a load by the force needed to lift it. When two wheels are used for a pulley system, a heavy load can be lifted with only a weak pull, but the load will move less far than the distance the rope is pulled.

YOU WILL NEED
● *2 pails* ● *empty reels used to hold wire (available at hardware stores)* ● *wire* ● *wire cutters* ● *scissors* ● *rope*

Adult help is advised for this experiment

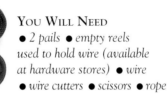

1 ASK AN ADULT to cut a long piece of wire. Thread this through the center of one reel. Use the wire cutters to bend one end of the wire around itself and the other end into a hook. Hang the pulley from a doorframe or the side of a bed.

2 TAKE A long piece of rope and tie one end around the handle of one pail. Use a firm knot. Put your load in the pail ready for lifting.

Tie the rope to the wire holding the top pulley and loop it around the bottom pulley first

Single pulley

3 THREAD the rope up over the pulley, and raise the pail to the height of your hips. Lift it by hand from hip height to shoulder height, and lower it. Now pull the rope to raise the pail the same distance. What is the mechanical advantage? Note how far you need to pull to raise the pail an equal distance.

Using a double pulley
You can make a double pulley by threading rope around two reels (see diagram above). You will need a lot more rope than you used for the single pulley. See how much effort you need to raise your pail and equipment. Is it more or less than before, and does the pail lift as high? (Answer on p. 186.)

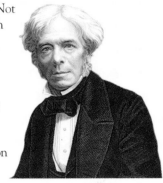

Thread the rope up round the top pulley and use this end to secure the load

Double pulley

What is zero?

THE NAME "ZERO" comes from the Latin *zephirum*, meaning empty, or blank. The symbol "0" originated in India; in A.D. 830 al-Khwarizmi (p. 68) explained the system of Indian numerals including the use of zero, but it was not translated for use in the West for another 400 years. Zero continued to puzzle scholars. Was it a number or a digit? If it stood for nothing, then surely it was nothing and did not need to be included. Leonardo of Pisa, or Fibonacci (1180–1250), in his book *Liber Abaci*, solved the puzzle. He said that zero can be used as a "place holder" to separate columns of figures. It can also represent a position on a scale. In temperature scales, zero degrees is a valid reading; it does not mean "no temperature."

Absolute zero

In winter, Lake Baikal in Siberia is one of the coldest places on Earth, reaching –76°F (–60°C). The lowest temperature, in theory, is absolute zero, equal to –459.69°F (–273.16°C). This may never be reached, although scientists have managed to cool atoms to a few millionths of a degree above absolute zero.

INVESTIGATING NUMBERS
Calculating with 0

Simple calculations of addition, subtraction, multiplication, and division using 0 will give interesting results, especially when you attempt to divide a number by 0.

ADDING 0
The number is unchanged.

$12 + 0 = 12$

SUBTRACTING 0
The number is unchanged.

$12 - 0 = 12$

MULTIPLYING BY 0
Multiplying a number by 0 always gives 0.

$12 \times 0 = 0$

DIVIDING BY 0
This will show an error on the calculator because 0 is the only number for which the operation of division makes no sense.

$12 \div 0 = E$ or error

Most calculators show the error symbol when division by 0 is attempted

EXPERIMENT
Sea level

Look up some lakes and inland seas in an atlas and find their heights above or below sea level. You should find some that are below sea level and others far higher above sea level than some mountains. Draw a graph (below) to show the differences.

YOU WILL NEED
● ruler ● felt-tip pens
● notepad ● graph paper

Make each bar a different color

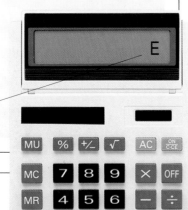

How to draw the graph
Jot down the levels of your chosen seas and lakes. Draw a graph with a vertical axis a little way in from the left edge, and the sea level line about a quarter of the way up the page. Write the feet in hundreds by the vertical axis, above and below the zero (sea level) line. Color in bars of equal width for the levels.

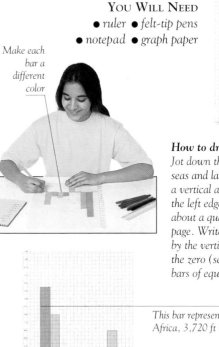

This bar represents Lake Victoria, East Africa, 3,720 ft (1,130 m) above sea level

This line represents zero altitude (sea level)

This bar represents the Dead Sea, Israel, 1,296 ft (395 m) below sea level

EXPERIMENT
Conversion chart

This chart can help you convert a temperature to either Fahrenheit or Celsius, using scales from 0°. Use a conversion table from a text book, agenda, or dictionary to find the correct conversions.

YOU WILL NEED
- *ruler*
- *selection of colored pens*
- *graph paper*

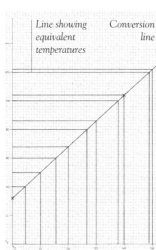

Line showing equivalent temperatures

Conversion line

1 ON THE GRAPH PAPER, mark the horizontal axis in 10°C divisions, from 0–50°C, and the vertical axis in 20°F divisions, from 0–140°F.

2 MARK THE POINTS 32°F/0°C and 122°F/50°C. Join the marks. You can now convert temperatures along this line.

Blasting off after the countdown
The countdown to the launch of a rocket may start several days before the event. It is computerized, but the final seconds may also be counted by voiceover. Ignition is usually at zero and this is designated T. Time before, or the hours and minutes to launch, is referred to as T–, and time after is T+. This is the launch of STS-31 mission, when the space shuttle Discovery launched the Hubble Space Telescope in April 1990.

EXPERIMENT
Making a thermometer

Temperature is merely a way of describing how hot or cold something is. There are different scales for measuring temperature; Fahrenheit and Celsius are the most common. Sometimes it may be necessary only to measure changes. You can make a thermometer, using colored water, to measure if the temperature is increasing or decreasing.

YOU WILL NEED
- *plastic bottle* ● *glass* ● *bowl*
- *pen* ● *modeling clay*
- *food coloring*
- *straws* ● *water*
- *ice*

1 SIPHON some of the water into the straw. Hold your finger over the end to stop water from running out. Put in a plug of clay to seal the top.

2 WRAP some clay around the top of the straw where it will be held in the neck of the bottle. Ask a friend to mark the water level on the straw.

3 INSERT THE STRAW into the bottle so that the end is just clear of the bottle's base. Seal the neck with the clay. Put the bottle in a bowl of hot water and watch the water in the straw. Put it in a bowl filled with ice. Does the water level go up or down according to the temperature? (Answer on p. 186.)

Positive and negative numbers

NUMBERS GREATER THAN ZERO are referred to as "positive," and those less than zero are called "negative." These numbers were known in ancient China. The Chinese had two sets of counting rods for calculating: red for positive numbers, and black for negative. Negative numbers were not often found outside China until the 16th century A.D., but now they are used worldwide. In the world of finance, when billions of stocks and shares change hands, the money is transferred electronically to bank accounts as positive or negative numbers; no notes or coins are transferred. In navigation, engineering, and science, the zero level is chosen for convenience. The zero level determines negative and positive measurements — the depth of an ocean or a canyon, for example, or the electrical charge of an atom.

❓ Gray matter

Arrange the digits 1, 2, 3, 4, and 5, using plus and minus signs (addition and subtraction), so that they total 111. Make sure you use every digit only once in your sum. Now, using the same rules, can you make 222 and 333? (Answers on p. 186.)

The person on the hopscotch moves two squares back from –1, which takes him to –3

Hopscotch

Positive and negative numbers can be demonstrated by this game of hopscotch for two players. One person stands on an arrangement of squares, and the other uses a dial to determine the moves. The dial is marked with six different moves, from – 3 to + 3. The players must work out the moves using mental arithmetic. You could decide the winner by taking turns to hop, with players leaving a marker on each square they land on until one person has left a marker on every square.

1 ONE player stands on 0. The other player moves the arrow between 1 and –1, to set the dial at 0.

2 SPIN the needle on the dial and wait until it has settled. The first player then hops the appropriate number of steps — in this case +2, or forward by 2.

3 ON THE NEXT SPIN, the arrow settles on –3. The hopscotcher works out the square to land on by doing the sum + 2 – 3, which is equal to –1, and hops back 3 squares.

Negative altitude

Submarines are used for surveillance and warfare, as well as for oceanographic research, exploration, and salvage operations. Whereas airplanes fly at a certain altitude, often fixed by air traffic control, divers and submarine captains need to know their depth below sea level — in other words, their negative altitude. An accurate measurement of depth also enables the crew to raise a periscope and view the surface when the craft is submerged.

EXPERIMENT
Tare weight

Tare weight is the weight of a container — such as a plastic food carton, a wooden crate, or a vehicle — before it is loaded with goods. Trucks are often weighed on a scale, and the tare weight is deducted from the total weight to find the weight of the load being carried. In this experiment, you can find the weight of something that might not hold still on the scales, such as a pet, by using your own weight as the tare weight.

YOU WILL NEED
- *bathroom scales*
- *object to weigh*

1 WEIGH YOURSELF. Make a note of your weight, and call that figure the tare weight. Remember to work in either metric or imperial; do not combine the two systems.

2 NOW WEIGH yourself with your pet. Note down the total weight of the two of you. Subtract your tare weight from the total to find the weight of your pet.

4 THE DIAL is spun again and settles on –2. The person on the hopscotch works out – 1 – 2 = – 3 before hopping to the –3 square.

Millions in debt

Wall Street in New York City is the financial capital of the United States. In 1929 the financial crash on Wall Street saw share values slump, leaving many ordinary people who had invested in shares in debt. In accounting, a debt is often written in brackets — ($3,500) or (£3,500) — to show that the number is negative (that it is owed).

Fractions and decimals

FRACTIONS HAVE BEEN USED for thousands of years. The Babylonian counting system, which used the number 60 as its base (pp. 46–47), is still used for telling time. It is simple to make fractions of an hour (60 minutes) — 1/2, 1/3, 1/4, 1/5, and 1/6 all give a whole number in minutes. In 1616 John Napier (p. 23) suggested that the decimal system, based on the number 10, would show any number, either greater or less than 1, by using a decimal point to separate the whole number from the fractional part. The digits to the left of the point represent the whole number (or 0), while those to the right show the fractional parts as so many tenths, hundredths, thousandths, and so on. Decimals are commonly used in measurement, in money, and on calculators; they allow numbers to be divided more precisely than fractions, although fractions sometimes give a clearer result.

EXPERIMENT

Fraction reckoner

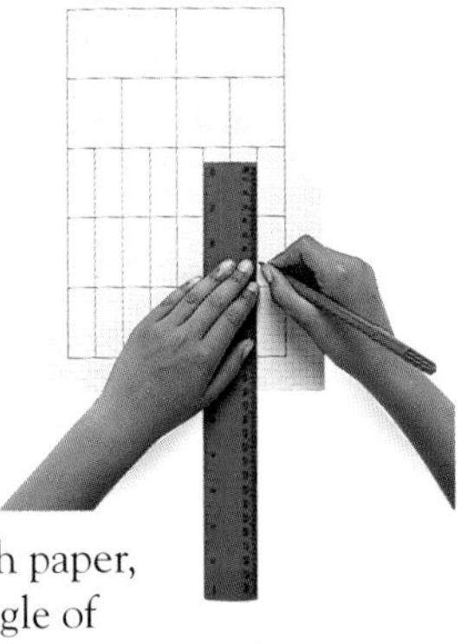

Equivalent fractions are those that are equal to one another, even though they have different forms. For instance 1/2 is equal to 2/4. This fraction reckoner will be useful as reference for simple calculations with fractions, to show which fractions have the same value.

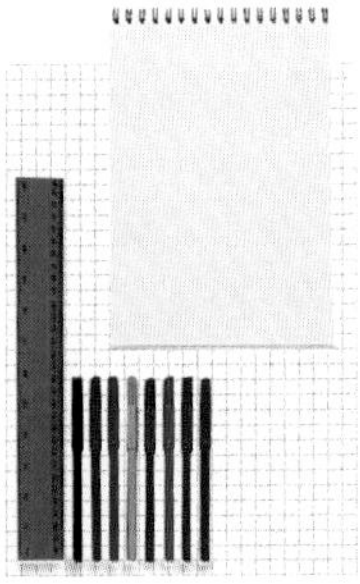

YOU WILL NEED
- *ruler*
- *notepad*
- *colored pens*
- *graph paper*

1 ON THE graph paper, draw a rectangle of 32×50 squares. Divide it into five rows 10 squares deep. Draw a line down the middle. In the second row, divide each half down the middle, to the bottom of the shape. Repeat the step for all rows.

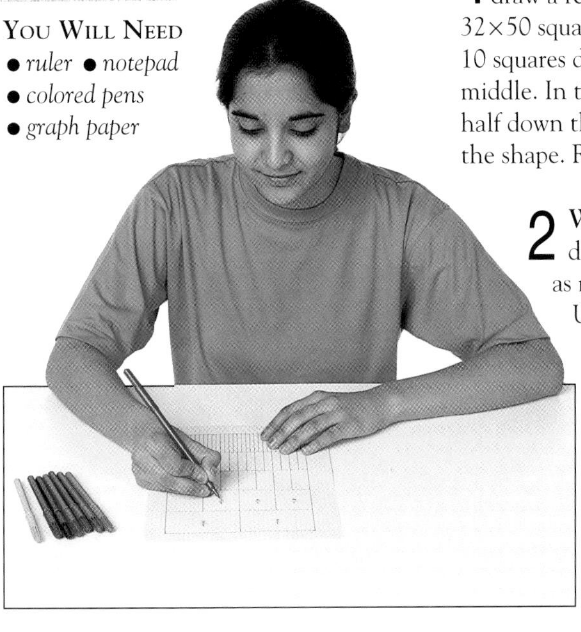

2 WHEN ALL the lines have been drawn, each row will have twice as many sections as the one above. Using a different color, write 1/2 in each section of the top row, and 1/4, 1/8, 1/16, 1/32 in each section of the lower rows.

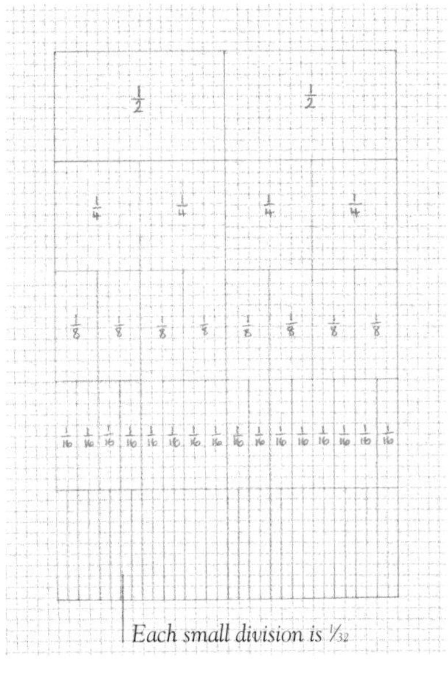
Each small division is 1/32

The fraction reckoner
The fraction reckoner can be used to show what the equivalent fraction is. If you put your finger on 1/4, you can see that it is the equivalent of 2/8, 4/16, and 8/32.

Carbon dating

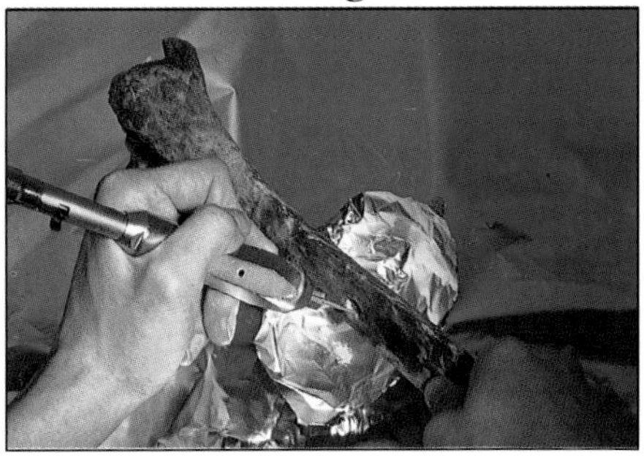

Radioactive carbon naturally loses half of its radioactivity every 5,570 years; this is its "half-life." Here, a sample of a prehistoric reindeer's bone is tested to discover what fraction of radioactive carbon remains and therefore how long ago the animal died.

Puzzle

Fill in the missing numbers to make a magic square (p. 26) of decimal numbers, in which each column, row, and diagonal adds up to the same number. (Answer on p. 186.)

2.00	0.25	1.50
	2.25	

DEMONSTRATION
Fraction of a book

Fractions are used every day to describe all sorts of things, such as how many pounds of apricots to buy or how much of an item has been used. To find how much of a book has been read, divide the number of pages read by the total number of pages.

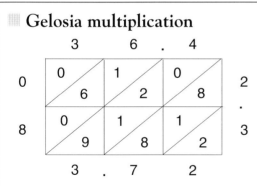

Simplifying the fraction
If the number of pages read is 300 and the total number of pages in the book is 400, divide 300 by 400: ³⁰⁰⁄₄₀₀. This fraction can be simplified very easily by canceling. Both the top number (numerator) and the bottom (denominator) can be divided by 100, just by removing the zeros. This gives the fraction ¼.

Turn to the last page to see the total number of pages in the book

🔢 Trick

1. Give two friends pieces of paper and pens. Ask each of them to pick a number between 1 and 99 and write the number secretly on a piece of paper. You are going to guess the numbers using a calculator.
2. Give the calculator to one of your friends and ask him or her to go through the following steps.
3. Key in the chosen number. Multiply by 2. Press =.
4. Add 4, and press =.
5. Multiply by 5, and press =.
6. Add 12, and press =.
7. Multiply by 10, and press =.
8. Without clearing the calculator, ask the first friend to pass it to the other person, who then adds his or her number to the total on the calculator.
9. Subtract 320 from the new total. Press =.
10. Divide by 100. Press =.
11. You look at the calculator. The digits to the left of the decimal point will show the number that your first friend chose, and those after the decimal point will be the number that your second friend chose. Ask both your friends to reveal the numbers they wrote down. They should match those on the calculator display.

Gelosia multiplication

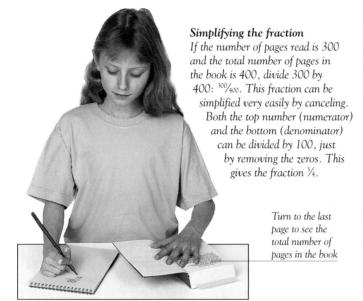

This technique was used in India from the 12th century onward. The diagram shows the sum 2.3×36.4. The number 36.4 is written along the top edge, and 2.3 is written down the right edge. Each digit at the top is multiplied in turn by each digit down the side, and the answers are written in the boxes, with tens in the top left of each box and units in the bottom right corner. Next, the numbers along each diagonal are added up, beginning at the bottom right corner, with the units written below the diagonals and the tens carried over to the next diagonal. Two lines are drawn from the decimal points to a diagonal that points to where the decimal point should go. The answer is 83.72.

DEMONSTRATION
Multiplying fractions

Fractions can be visualized more clearly than decimals; one half is easier to imagine than 0.5. Some fractions can also be written and used more simply. For example, when 1 is divided by 3, the fraction is ⅓, but the decimal is 0.3̇; the dot over the last digit shows that the 3 is repeated endlessly. If you multiply ⅓ by 6, the answer is exactly 2 (⅓ × 6 = 2). But because 0.3̇ is endless, it is very hard to use in calculations. If reduced to, say, 0.333 and multiplied by 6, it gives only 1.998. The cake below shows both what a fraction looks like and what happens if a fraction is multiplied by itself.

Dividing up a cake
First the cake is divided into two equal parts by being cut down the center. There are now two halves. If each of these halves is divided into two equal parts, then there
will be four parts, or four quarters. You can see that a half of a half (½×½) is equal to one quarter (¼). The answer is smaller than the original fractions.

Estimating

ALTHOUGH MATHEMATICS is a precise discipline, in practice we often need only approximate answers. For example, companies packaging goods estimate the amount of those goods consumers will likely use, and economists estimate financial trends. Someone about to cross a street estimates the speed of oncoming cars so that they know whether or not to wait before crossing. Often we instinctively round off a figure to an approximate value. You may say that you are 12 years old, rather than 12¼, or that you will be home in about 20 minutes, not 18 minutes and 42 seconds. Making mistakes with a calculator is easy; to spot them you should first make an estimate.

Estimating support for a cause

The police often have to estimate the size of crowds, such as this political demonstration in Germany. They may produce a figure rounded up or down to the nearest thousand. The organizers of the demonstration might give a larger figure to emphasize the strength of their support.

EXPERIMENT
Estimating by weight

Often, when working with large numbers or quantities, a sample is first taken to find an average or mean (p. 83). The average is then used mathematically to estimate the answer. Peanuts, for example, are not all the same size and weight, but if you take a large enough sample to provide an estimate of the weight of an average peanut, you can then calculate how many peanuts there might be in a bag of a particular weight. Retailers usually quote only an approximate number of items in a packet of a precise weight.

YOU WILL NEED
● *calculator*
● *bag of peanuts*
● *notepad* ● *pen*
● *scales*

1 WEIGH THE whole bag of peanuts. Make a note of the weight. The bag used here weighed 30 oz (840 g). See if the manufacturer's weight corresponds to the weight that you measure.

The whole bag of peanuts is weighed

Count the nuts in the weighed sample

Weigh enough peanuts to get a minimum reading

2 WEIGH OUT a handful of peanuts. Choose a round figure such as 1 oz or 30 g and take away or add enough peanuts to make up this weight. You need to weigh a large enough amount so there is a reading on the scales, but not so many that you are counting into the hundreds.

3 TIP THE PEANUTS out in front of you. Count them and write down the number in the sample. To estimate the total number of peanuts in the bag, divide the weight of the bag by the weight of the sample, and multiply the result by the number of nuts in the sample.

Note the weight of the sample

EXPERIMENT
Pure guesswork

Try this simple test of judgment. Fill a jug with colored water. Find two different-shaped containers and a bowl, and guess how many measures of water from each container will fill the bowl. Pour measures from the containers into the bowl to see how accurate you were. With experience, your guesswork will get better.

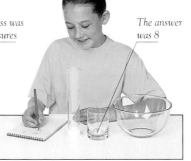

The guess was 14 measures

The answer was 8

YOU WILL NEED
- *jug of water* • *2 containers of different shapes*
- *bowl* • *notepad* • *pen* • *food coloring*

1 WRITE DOWN your guess for the number of times the first container will fill the bowl. Fill the container, and pour measures into the bowl until it is full.

2 NOTE THE NUMBER of measures used to fill the bowl. Compare this figure with your guess. Repeat using the other container. See if your friends can do any better.

EXPERIMENT
Taking a sample section

A quadrat is a sample unit of land used by biologists and other scientists to study the growing patterns of plants and other living things in a specified area. Studies are made of small parts of the area, and from this the scientist estimates general patterns for a larger area (such as a field or shoreline). Using a frame to mark out the sample area, you can plot ecological maps of plant life and see how much this changes over time. Here, the number of daisies in a patch of grass is being estimated.

YOU WILL NEED
- *ruler* • *pencil* • *scissors*
- *notepad* • *thick poster board*

1 TO MAKE A FRAME, measure the length of your foot, and draw a square with sides of that size. Draw a larger square about 1½ in (4 cm) out from the first square. Cut out the square along the lines to make a frame.

2 TAKE THE FRAME to a yard or a park and lay it anywhere on the grass. Count the number of daisies in the square, and write this number down. Repeat this to get an average number (p. 83).

3 PACE THE edges of a patch of the grass, and note the length and width in foot-lengths. Work out the area (pp. 98–99). Because the sample is equal to one square unit, multiply the number of daisies in the sample by the units of area of your patch of grass to find the approximate number of daisies in the patch.

OTHER SORTS OF NUMBERS

Sᴏᴍᴇ ɴᴜᴍʙᴇʀꜱ ʜᴀᴠᴇ special qualities. Far from being mere curiosities these numbers can be used for tasks such as coding secret messages and describing how electrical circuits work. The study of the nature and properties of numbers is called number theory. This area of mathematics has existed for nearly 2,500 years, and is still being developed.

Mathematicians call the numbers that we use for counting "natural numbers." These can be used to count collections of objects such as sheep in a field, pebbles on a beach, or stars in the sky.

The followers of the Greek philosopher Pythagoras (p. 124) used geometry, the study of shapes and patterns, to develop many theories about natural numbers that are still in use today. The Pythagoreans were fascinated by squares and other geometric shapes, and studied the square roots of natural numbers (p. 41). They found that some natural numbers, such as 4, 9, 16, 25, and 36, had square roots that were also natural numbers, and that others had square roots that could be written as fractions.

A few natural numbers, though, such as 2, 7, and 11, had square roots that could not be written as fractions. The square roots of such

The mad mathematician
This drawing of a group of lunatics by William Hogarth (1697–1764) shows a mathematician scribbling on the wall in the background. Ordinary people often regarded mathematicians as eccentric or even mad.

numbers are "irrational numbers" – this means that they cannot be written exactly as a ratio (p. 53). Bearing in mind that a decimal number is actually a fraction ($0.001 = \frac{1}{1000}$) then, for example, a calculator will give $\sqrt{2}$ as a number such as 1.4142135, with as many digits after the decimal point as it can display. The exact value would have an infinite number of digits after the decimal point. Irrational numbers include pi (p. 134) and the Golden Section (p. 58).

■ Enormity

As mathematics has evolved, increasingly large numbers have been needed to give shape to ideas. The Greeks were aware of such numbers, but had no simple way of writing

them down. Archimedes (p. 18), in his treatise *The Sand Reckoner*, suggested a system based on the myriad, which would be equal to 10,000 in base 10. The largest number in this system, if written in base 10, would be 1 followed by 80 million billion zeros. Archimedes then estimated the number of grains of sand that would be needed to fill the Universe as 10^{51}.

■ Perfect and prime numbers

Two other types of numbers at the heart of number theory are "perfect" and "prime" numbers. Perfect numbers, like irrational numbers, were discovered by the disciples of Pythagoras. A perfect number is the sum of all the numbers that divide into it exactly, including 1 but excluding the perfect number itself; 28, for example, is perfect. It is divisible by 1, 2, 4, 7, and 14, and $1 + 2 + 4 + 7 + 14 = 28$.

Prime numbers are numbers that are exactly divisible only by themselves and 1. Primes were

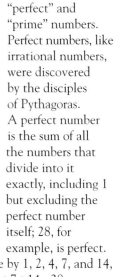

Alien numbers
Base 10 arose because of humans' natural tendency to count on their fingers. If alien life forms exist, they might count in bases that originate in the same way. These "alien" fingers show base 6.

Adelard of Bath
An English-born scholar, Adelard traveled extensively throughout Europe in the late 11th century. He obtained an Arab translation of Euclid's Elements (p. 114) and translated them into Latin, so helping to reintroduce Euclid to Europe. His translation was the chief mathematical textbook in Europe for many centuries.

Cardboard computing
*This database (p. 47)
holds information on guests
for a party. It works like
the punch cards in early
computers. Information is
shown in binary code, as a
series of holes or open
slots on each card.*

first discussed 200 years
after Pythagoras' death,
by the Greek philosopher Euclid
(p. 114). Euclid wrote no less
than 13 books on geometry,
arithmetic, and number theory.
He called his books *Elements*.

The smallest prime, and the
only one that is an even number,
is 2 (1 is not viewed as prime).
Euclid proved that an infinite
number of primes exists. In
addition, he developed a theorem
now known as The Fundamental
Theorem of Arithmetic. This
states that all natural numbers
can be formed by multiplying a
unique set of prime numbers,
unless they are prime themselves.

Mathematicians have never
lost their fascination with prime
and perfect numbers. It is now
known to what degree prime
numbers become scarcer, and
so harder to find, among larger
numbers. There are still some
questions about primes that
remain unanswered. For example,
can every even number greater
than 2 be written as the sum
of two primes? This has never
been proven formally, but a
famous statement, known as the
"Goldbach conjecture" states
that the answer is "yes."

Because large
primes are so hard
to find, they are
often used by
institutions, such
as hospitals and
banks, when they
need to send
confidential
messages. The
senders can
scramble information using two
large primes known only to
themselves and the intended
receiver — rather like personal
identification numbers, or PINs,
in banking (p. 45). Because only
parts of the information are
known to any one person in the
chain, this makes it practically
impossible for anyone but the
receiver to decipher it.

In 1995, for the first time ever,
an American mathematician
claimed a legal patent on two
large prime numbers for use in
coding and security for computer
systems. In 1992, scientists
in Britain discovered a new
prime number by using a Cray-2
supercomputer. The prime they
found is $2^{756,839}-1$. If the number
were written in full, it would be
227,832 digits long!

Bases

The properties of numbers are
independent of the base in which
they are counted. We count in
base 10 but mathematics has
also been developed by people
who work in other bases. The
Babylonians, for example,
counted in bases 60 and 360.

A very important base used today
is base 2, comprising the digits 1
and 0 (p. 46). Base 2 numbers are
the easiest numbers to store and
process in electronic circuits.
A voltage in a circuit can be
switched off to represent a 0 or
on to represent a 1. All processes
inside computers are carried out
with base 2, or binary, numbers.

Because numbers keep their
special properties when they are
written in binary code, computers
can be employed to help people
study the properties of numbers
further. However, computers
are not the best tools to resolve
problems such as the Goldbach

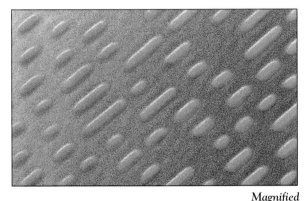

conjecture. The proof or disproof
of such theories will demand the
genius of mathematicians who
are as creative as Euclid, Euler
(p. 164), or Pythagoras.

Arranging numbers

Patterns are everywhere in
numbers. Two common types
are series and sequences (p. 48).
Another is Pascal's triangle. It
was already known in China in
the 14th century, and was studied
by Blaise Pascal (p. 79) in
the 17th century. He saw
that the triangle showed
several patterns at once,
such as square numbers,
triangular numbers, and
the Fibonacci sequence.
The ability to see the patterns
that link numbers has allowed
great advances in mathematical
and scientific thinking.

*Magnified
compact disc*
*The sound information
on a compact disc is
arranged on a spiral
track consisting of
about one billion tiny
pits of different lengths,
with smooth areas
between them. A laser
in the CD player shoots
a beam of light at this
track. The reflected
light is read in binary
code; each pit is a 0
and each flat part a 1.*

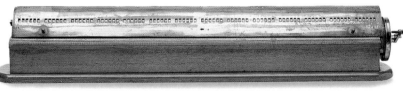

Astronomical calculator
*The enormous distances between planets and stars can be expressed only by using
very large numbers. Astronomers in the 19th century worked with huge numbers,
and their assistants used hand-cranked machines such as this one to calculate and
interpret the figures. This machine can show figures of up to 42 places.*

Using powers

IF A NUMBER is written to a power, that number is multiplied by itself. So 7 to the power of 4, or 7^4, $= 7 \times 7 \times 7 \times 7 = 2,401$. Mathematicians have known of the concept of powers for centuries, but the rules for their use were formalized only in the 14th century. Since that time, accurate analysis of numbers, for example in Cartesian geometry (p. 74), has been possible, and the use of powers is now fundamental in all forms of mathematics.

INVESTIGATING NUMBERS
Odd numbers and powers

When consecutive odd numbers are added together, a pattern emerges that provides a quick way of calculating the sum of these numbers.

1 Add the first two odd numbers; the answer is an even number.

$$1 + 3 = 4$$

2 The number 4 is 2×2, which can also be written as 2^2 or 2 squared.

$$1 + 3 = 4$$
$$= 2 \times 2$$
$$= 2^2$$

3 Add the first three odd numbers. What do you get? Add the first six odd numbers. Can you see a quick way to the answer? (Answer on p. 186.)

$$1 + 3 + 5 = 9$$
$$= 3 \times 3$$
$$= 3^2$$

Newton's workings

This calculation was written by Sir Isaac Newton (p. 71) in 1665, and shows his attempt to work out the area under a hyperbola (a type of curve, p. 140). In it he calculates the answer to 55 places and uses powers to write the numbers in a manageable form. They appear at the ends of the lines down the right-hand side.

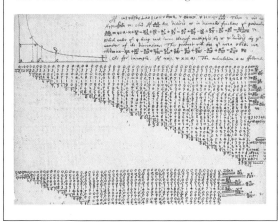

Puzzle

Find the factors (p. 45) of 25, 100, 144, 169. What is special about these numbers? (Answer on p. 186.)

EXPERIMENT
Understanding powers

A very simple exercise — folding a sheet of paper — can help you understand how the power of a number increases successively. You will find that, no matter how big your piece of paper or how thin it is, there is a limit to the number of creases you can make. As you fold the paper, you will divide it into an increasing number of sections. These sections are related to the number of creases. Do this experiment with the largest sheet of paper you can find. You will be able to find out how many times the paper has to be folded to give 64 or 128 sections.

YOU WILL NEED
- *large piece of paper*

1 LAY THE PIECE of paper on the table in front of you. Fold it exactly in half, either vertically or horizontally. Sharpen the crease with your fingernails so that you will be able to see it clearly.

2 UNFOLD THE PAPER. You have now divided the paper into two parts with one central crease. You could write this as 2^1 (2 to the power 1) — this signifies two parts and one crease.

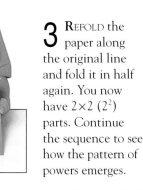

3 REFOLD the paper along the original line and fold it in half again. You now have 2×2 (2^2) parts. Continue the sequence to see how the pattern of powers emerges.

Groups of numbers

Numbers can be grouped according to their powers, and tables of powers, such as those for square and cubic numbers, are used by mathematicians. The relationships between these numbers have preoccupied number theorists for centuries. One notable problem was devised in 1770 by a British mathematician, Edward Waring (1734–98), who wrote that "every integral number is either a cube itself or the sum of a number of cubes." Test this idea for yourself and see if Waring was right. (An integral number is any positive or negative whole number, pp. 32–33.) Choose a number and see if you can add cubes of numbers (other than 1 or –1) to equal that number. For example, $75 = 4^3 + 3^3 + (-2)^3 + (-2)^3$. Use a calculator to help.

Linear numbers

A number, say 2, is linear if it is raised to the power of 1. It can be written as 2^1. A number to the power 1 is equal to itself, so the power sign is usually dropped. Powers can be thought of in terms of dimensions, with linear numbers being one-dimensional. This can be shown with marbles arranged in a straight line.

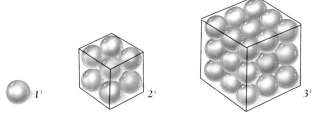

1^1 2^1 3^1 4^1 5^1

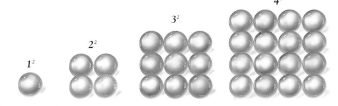

1^2 2^2 3^2 4^2

Square numbers

The square of a number, say 3^2 (3 to the power 2), is that number multiplied by itself: $3 \times 3 = 9$. It is also the number of units in a square grid having sides of a given number. If we show squares using marbles, 1^2 has 1 marble, 2^2 has 4 marbles, 3^2 has 9 marbles, and so on. Numbers that are multiplied by themselves to give squares are called square roots. For example, the square root of 9 (written as $\sqrt{9}$) = 3.

1^3 2^3 3^3

Cubic numbers

A cubic number is a number that is multiplied by itself, and then by itself again. For example, the cubic number 4^3 (4 to the power 3) equals $4 \times 4 \times 4 = 64$. In the pictures above, cubic numbers are shown as cubes of marbles, with all their sides the same length. The number that is multiplied to make a cubic number is referred to as a cube root. The cube root of 64, for example, is 4. This is written as $\sqrt[3]{64}$.

INVESTIGATING NUMBERS

Calculating with powers

There are quick ways to calculate with powers, but the operations are not immediately obvious. The calculations can be written in different ways to show why the methods are used.

1 To multiply the same number raised to powers, add the powers and convert the result.

$5^2 \times 5^3 = 5^{(2+3)}$
$5^5 = 3,125$

2 The sum can be illustrated another way if the whole sum is written out in full.

$(5 \times 5) \times (5 \times 5 \times 5)$
$= 5 \times 5 \times 5 \times 5 \times 5$
$= 5^5$
$= 3,125$

DIVIDING A NUMBER
To divide a number raised to certain powers, subtract the powers.

$5^3 \div 5^2 = 5^{(3-2)} = 5^1$
$(5 \times 5 \times 5) \div (5 \times 5)$
$= 125 \div 25$
$= 5$

TO THE POWER OF 0
Use this method to show that any number to the power of 0 is equal to 1.

$5^2 \div 5^2 = 5^{(2-2)} = 5^0$
$(5 \times 5) \div (5 \times 5) = 25 \div 25 = 1$
therefore $5^0 = 1$

♣ Puzzle

Use your calculator to see the patterns that emerge when you square numbers made up solely of the digits 1 or 3. For example:

$1^2 = 1 \times 1 = 1$ $3^2 = 3 \times 3 = 9$
$11^2 = 11 \times 11 = 121$ $33^2 = 33 \times 33 = 1,089$

Continue with these calculations and see what the pattern is. Using the patterns, can you guess the squares of 1111111 and 3333333? (Answer on p. 186.)

INVESTIGATING NUMBERS

Squaring fives

This quick technique will enable you to calculate the square of any number less than 100 that ends with a 5.

1 To square 65, ignore the 5. Multiply the first digit, which is 6, by itself after adding 1 — in this case, 7.

65^2

$6 \times 7 = 42$

2 Write this number down and put 25 after it to get the correct answer.

$4,225$ therefore
$65^2 = 65 \times 65 = 4,225$

Big numbers

VERY LARGE NUMBERS are used in many fields, from mathematics to engineering, technology, astronomy, and geography. Infinity, an amount with no limit, is the largest of all. If numbers are too big to write easily, standard numbers called powers (p. 40) are used to express them in a more convenient way. In measurement, large numbers are usually put in simpler terms. For instance, rather than saying you walk 126,720 inches, you would use the equivalent distance of 2 miles. In metric units, "kilo-" (×1,000), "mega-" (×1,000,000), and "giga-" (×1,000,000,000) are a convenient way to describe big quantities.

Light-years

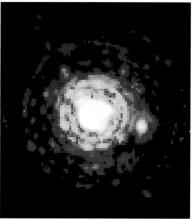

The distance between this star, Gliese 623.a, and the bright spot to the right is 200 million miles (320 million km). Distances in space are measured in light-years. One light-year is equal to the distance traveled by a ray of light in a year: 6 million million miles, or 9.5×10^{12} km (p. 40), which is about 9.5 million million km.

Electronic resistor

The value of this electronic resistor is marked with a system of colored stripes. This is a way of writing large numbers on a tiny object. The first two stripes show a number, and the third stands for a multiplier expressed as a power of 10. The stripes here show 1 (brown), 0 (black), followed by 10 to the power 2 (red), that is 10×10^2 or 1,000 ohms.

EXPERIMENT
Time line of life on Earth

To make this time line you will need to use an encyclopedia to research the development of life on Earth. Starting from 3,500 million years ago, find the approximate dates for the emergence of single-celled organisms, flowers, fish, amphibians, birds, forests, dinosaurs, mammals, and humans. With this data you can make a time line to hang on your wall.

YOU WILL NEED
- ruler ● colored pencils
- pens ● adhesive tape ● scissors
- paper (preferably graph paper)

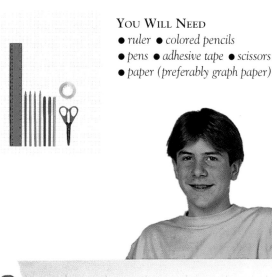

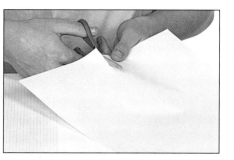

1 CUT SOME paper sheets in half lengthwise and join the strips with adhesive tape to make one long strip about 11 ft 4 in (3.5 m) long and 6 in (15 cm) wide. Draw a line along the base, 1 in (2.5 cm) up from the bottom edge and ½ in (13 mm) from each side.

2 MARK OFF THE LINE in 4-in (10-cm) sections. The scale here is ¼ in (6 mm) for 5 million years. Using your data, draw vertical lines for important events, such as the emergence of the amphibians (410 million years, or 16 in/ 41 cm along the time line).

Draw vertical lines marking divisions of 10 million years

Write the number of years represented under each vertical line

Standard form

2.34   11

Very large numbers are not easy to work with when written down in full, so standard forms are used as a convenient way of expressing them. A large number such as 234,000,000,000 is written as a decimal number between 1 and 10, multiplied by 10 to the appropriate power (p. 40); for example, 2.34×10^{11}. Standard forms are available on many calculators (see above), although some may show an "error," or E, if the numbers are too large for them to handle.

Gray matter

Using a calculator, work out how many seconds you have been alive, including the 24 hours of the day you were born and the 24 hours of the day of your calculation. Estimate first and see how accurate you were. Write down all the calculations and check your answers on p. 186. Remember that there are also leap years (p. 45).

EXPERIMENT
What is infinity?

Infinity is a quantity larger than anything that is fixed; it has special significance in areas such as number theory, algebra, and geometry. For instance, the list of whole numbers is 1, 2, 3, 4, 5, . . . Because it is always possible to add 1 to any whole number, the list is never-ending. This experiment gives an impression of infinity as something that never ends. Infinity is sometimes represented by the symbol "∞."

The only limit is the reflections that your eyes can distinguish clearly

YOU WILL NEED
● *two mirrors, one larger than the other*

A multiple you
Hold one mirror and ask a friend to hold the other. Stand so you can see yourself in his mirror through yours. If the mirrors are perfectly straight and the angle is correct, there will be more than 12 reflections.

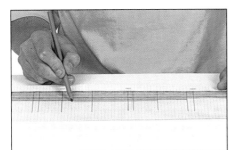

3 WHEN YOU HAVE ruled the vertical lines up to the arrival of humans, color in different bands from the right-hand end of the line to the appropriate vertical line to show how long these creatures have been on Earth. Some, like the dinosaurs, will have died out.

Draw extra vertical lines to show the emergence of each species

Our short life span
Following the history of the Earth with this time line, it is humbling to see that humans have inhabited it for such a short space of time.

Make each line one row of squares high, or ¼ in (6 mm)

Special numbers

SPECIAL NUMBERS ARE NUMBERS with qualities that make them different. These numbers include primes, perfect numbers, square numbers (p. 41), and triangular numbers. The special properties of these numbers may allow them to be written as a formula, say n^2. Some properties allow huge numbers to be written simply so that they give a great deal of information that has already been worked out. For example, a particular Fermat prime known as F_{1945} is actually a number with more digits than the number of particles in the entire universe and so cannot be written down in full. Special numbers are now used in computer security to encode information and keep it secret.

Maria Goeppert Mayer

Maria Mayer (1906–72) shared the Nobel Prize for physics in 1963 for her work on the pattern of "magic numbers" in atomic nuclei. She found that nuclei that have 2, 8, 20, 28, 50, 82, and 126 protons or neutrons are stable.

The numbers are determined by the physical properties of the atoms. Mayer first trained as a mathematician before turning to physics, and she was one of the few women to be involved in atomic physics at the time.

EXPERIMENT
Eratosthenes' sieve

Eratosthenes was a Greek astronomer who lived around 200 B.C. He calculated the circumference of the Earth (p. 135) from simple observations of the Sun's rays shining down a deep well. Eratosthenes' sieve is a systematic way of isolating prime numbers. A prime number is divisible only by itself and by 1, although the number 1 is not a prime number.

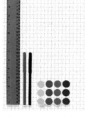

YOU WILL NEED
- *ruler* ● *pens*
- *colored counters*
- *graph paper*

1 USING A BLACK PEN, draw a large square of 50×50 small squares. Divide this into 10×10 squares. With a red pen, write in the numbers 1–100, starting in the top left corner.

2 PLACE A BLUE counter on 1 and then place red counters on every multiple of 2, but not 2 itself. Put yellow ones on multiples of 3, but not 3; blue on multiples of 5, but not 5; green on multiples of 7, 11, and so on. The uncovered numbers are prime numbers.

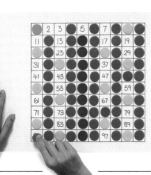

Gray matter

It is usually difficult to find any pattern to prime numbers, but here are some that follow the formula $n^2 + n + 17$.

$$17 \quad 19 \quad 23 \quad 29 \quad 37 \quad 47 \quad 59$$

What are the next five numbers in this series? Are they all primes? Is the formula true for values of n greater than 15? (Answer on p. 186.)

INVESTIGATING NUMBERS
Finding perfect numbers

A perfect number is one that is equal to the sum of its factors, including 1 but excluding itself. (Factors are smaller numbers that divide a number exactly.) For example, $6 = 1 + 2 + 3$. You can find all even perfect numbers by applying this formula: $2^{n-1}(2^n - 1)$, where n is a prime number.

1 In the formula, replace n with the prime number 3.

$$2^{3-1}(2^3 - 1) = 2^2(2^3 - 1)$$
$$= 2^2(8 - 1) = 2^2(7) = 4(7) = 28$$

2 Add the factors of 28 together (excluding 28).

$$1 + 2 + 14 + 7 + 4 = 28$$

3 Now use the power 5, the next prime number.

$$2^{5-1}(2^5 - 1) = 16(31) = 496$$

4 Use a calculator to find the factors of 496 (excluding 496) to show that this is a perfect number.

$$1 + 2 + 248 + 4 + 124 +$$
$$8 + 62 + 16 + 31 = 496$$

EXPERIMENT
Factor finder

YOU WILL NEED
- *colored pens*
- *graph paper*
- *ruler*

A factor of a number is any number (including itself and 1) that divides into it exactly. Our chart covers the numbers 1–25 and is useful as a reference for finding their factors. You need to work out the factors for all these numbers yourself. The numbers up the side of the chart represent the possible factors.

1 ON THE PAPER, draw lines along the base and up the left side for the axes. Write the numbers 1–25 along each axis, with 1 in the bottom left corner.

2 STARTING AT THE BOTTOM left, fill in 1–25 on the diagonal. Write 1 in each square of row 1, and the other factors where the factor row meets the number column.

3 TO FIND THE FACTORS of a number, find that number along the bottom row; the factors will be those numbers written in a column above.

EXPERIMENT
Keeping secure

Security of information and privacy are vital aspects in private and public life. Codes made up of numbers are used to prevent crime. For example, personal identification numbers are used to gain entry into a bank account. And secret codes are used to create ciphers in computer systems. Even though a word might be used, the letters are converted into numbers. Only the receiver has the code to decipher the message, keeping the information secret.

YOU WILL NEED
- *notepad* • *calculator*
- *pen* • *combination lock or padlock*

Setting the combination
Use a personal number to set a code on your lock. Think of a three-letter word and encode it. Attribute a number to each letter; the simplest is 1 = A, 2 = B, and so on. A word is usually easier to remember than a group of numbers. Set the combination on the lock.

INVESTIGATING NUMBERS
Modular arithmetic

This type of arithmetic is a way of writing numbers by dividing and looking at numbers left over. It is useful in cyclic situations, such as days of the week or the time in 12- and 24-hour clocks.

LEAP YEAR – MODULO 4
Divide the number by 4; if it has no remainder, the year is a leap year.

$1996 \div 4 = 499$
1996 is a leap year

THE TIME – MODULO 12
Divide the time on a 24-hour clock by 12. The number left over is time on a 12-hour clock.

$2200 \div 12$
$= 1 \quad 10$ left over
$= 10$ o'clock

BOOK NUMBERS – MODULO 11
1 Modulo 11 can be used to see if the codes on books' ISBNs (International Book Standard Numbers) are correct.

ISBN
0 1 4 0 5 1 1 1 9 9

2 The first number is multiplied by 10, the next number by 9, the next by 8, and so on until the last number is multiplied by 1.

$0 + 9 + 32 +$
$0 + 30 + 5 +$
$4 + 3 + 18 + 9 = 110$

3 The numbers are added up and divided by 11. If the ISBN is correct, there will be no number left over.

$110 \div 11 = 10$
and nothing left over

Number bases

WHEN DIGITS ARE WRITTEN in a row to represent a number, each place in the row (p. 30) has a different value. Different civilizations have used different number bases. The imperial system of weights and measures uses 12 inches to a foot and 16 ounces to a pound. Today, the most common base is 10 — units, tens, and hundreds — but there are other bases too. For instance, we tell the time in base 60, with 60 seconds to a minute and 60 minutes to an hour. And computers, which operate digitally, use the binary, or base 2, system, in which numbers are represented by 0's and 1's. Messages fed in are converted into binary code and stored, or are passed on as electric signals, with the current being switched on for 1 and off for 0.

DEMONSTRATION
Alien fingers

We use base 10 because we have 10 digits on our hands. In this demonstration you can work in base 10 and a friend can pretend to be an alien with six digits — she uses base 6. Try working with other bases too.

INVESTIGATING NUMBERS
Bases 10, 2, 60, and 12

The number 105 is written in base 10 (the denary system). It looks different written in other number bases. In all bases the right-hand column represents the base number (n) to the power 0 (p. 41), the next column to the left represents n^1, the next, n^2, and so on.

	n^6	n^5	n^4	n^3	n^2	n^1	n^0
BASE 10 The number is formed of: $1\times100, 0\times10, 5\times1$.					1	0	5
BASE 2 The number is made up like this: 1×64, $1\times32, 0\times16, 1\times8$, $0\times4, 0\times2, 1\times1$.	64 \newline 1	32 \newline 1	16 \newline 0	8 \newline 1	4 \newline 0	2 \newline 0	1 \newline 1
BASE 60 The number is formed like this: 0×3600, $1\times60, 45\times1$.				3600 \newline 0		60 \newline 1	1 \newline 45
BASE 12 The number is formed like this: 0×144, $8\times12, 9\times1$.				144 \newline 0		12 \newline 8	1 \newline 9

6 digits

6 digits

1 digit

Both are communicating numbers, but in different bases

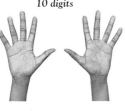

10 digits

3 digits

1 IN BASE 10, 13 is made up from a full set of 10 fingers, plus another three fingers: this is one ten and three units.

2 THE ALIEN HAS ONLY SIX fingers. By looking at the two sets of hands you can see exactly what each base looks like. It may help if the "alien" counts aloud up to the required number because she will need to use her fingers three times (right).

3 IN BASE 6, 13 is 2 sixes plus 1 unit. If the number were written out in base 6, it would be written as "21."

EXPERIMENT
Making a computer

Information can be stored on computers in the form of a database. This contains "fields" that represent categories of data. Information is processed in binary code. The simple database below uses cards to give category information about party guests, such as boy, likes music, likes dancing, is allowed out at night. The data is stored as a closed hole (0) for yes or a slot (1) for no. Before the arrival of magnetic disks, many computers used punched cards like these to store data.

Adult help is advised for this experiment

1 DRAW SYMBOLS on paper for eight categories. Copy each several times. Color and cut them out. Stick them to data cards, 1 in (2.5 cm) apart, in the same order on each card.

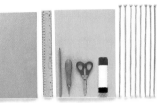

2 GLUE EACH friend's picture on a card. Ask an adult to make holes directly above the symbols that apply to that person, and cut slots for those that do not.

YOU WILL NEED
- *covered cereal box* ● *ruler* ● *pencil* ● *bradawl*
- *scissors* ● *8 knitting needles* ● *glue* ● *pen* ● *data cards*

3 MAKE ANOTHER COPY of the eight symbols. Stick them onto small rectangles of card to form the markers. Ask an adult to pierce each marker and push it onto a knitting needle.

4 CUT A SLIT ACROSS the base of the box at the front, and remove the top. Ask an adult to make eight holes at the top of the box for the needles. Use one card as a template for the holes.

5 HOLD ALL the cards together. Push needles through the box and the cards, in the same order as the symbols.

6 TO CHOOSE who to invite to your party, decide on a few categories. Pull out the needles for these categories, and leave all the other ones in place.

The markers on the needles correspond to the parts of each card with holes rather than slots

7 BY THE TIME you have removed all the relevant needles, the cards for friends that fit the chosen categories will have fallen to the bottom of the box. You have your party list.

The machine has selected people belonging to all the relevant categories

■ DISCOVERY ■
Gottfried Wilhelm Leibniz

Gottfried Leibniz (1646–1716) was born in Leipzig in Germany. In mathematics he was largely self-taught and is perhaps best known for his work on calculus in 1675, which he developed independently of Sir Isaac Newton (p. 71). He also built an early calculating machine. In 1679 Leibniz perfected the binary system, now widely used in computing and electronics. He saw mystery in the system, equating 1 with God and 0 with nothingness.

Series and sequences

PATTERNS ARE EVERYWHERE in numbers. The identification of new patterns and their uses has intrigued mathematicians for centuries. A "sequence" is a list of numbers that follows a certain pattern; the most simple is 1, 2, 3, 4, 5 . . ., where the numbers increase by 1 every time. A "series" is different: it is a sum of the numbers in a sequence. For instance, the sequence 1, ½, ¼, ⅛, ¹⁄₁₆ . . . may have a series 1 + ½ + ¼ + ⅛ . . . . The more numbers are added to this sequence, the closer this series comes to a value of 2. Much early work on series and sequences had only theoretical value, but the early "pure" study has produced practical applications much later, particularly in science and engineering. For example, in biology the division of cells follows the sequence 1, 2, 4, 8, 16, and so on (the power of 2 is increased by 1 each time: 2^0, 2^1, 2^2, 2^3, 2^4).

EXPERIMENT
Pascal's triangle

YOU WILL NEED
● *calculator* ● *graph paper or paper ruled like a grid* ● *pens*

Frenchman Blaise Pascal (1623–62) was a mathematical prodigy. The arithmetical triangle shown here had been known for 600 years, but Pascal discovered that many of its properties were linked to special sequences and series. The triangle in this experiment shows the Fibonacci sequence, square numbers (p. 41), and number sequences within algebraic formulas (p. 70). It also has links with probability (p. 80).

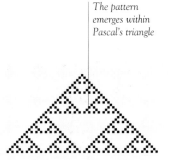

1 STARTING AT the top center square of graph paper (the apex), write the number 1. On the diagonal in both directions, continue writing the number 1 to mark both sides of the triangle.

The pattern emerges within Pascal's triangle

Triangular numbers

Natural or linear numbers

To find the number of ways of selecting 3 dishes from a total menu of 7, go to the 4th number (3 + 1) on the 7th row; the answer is 35

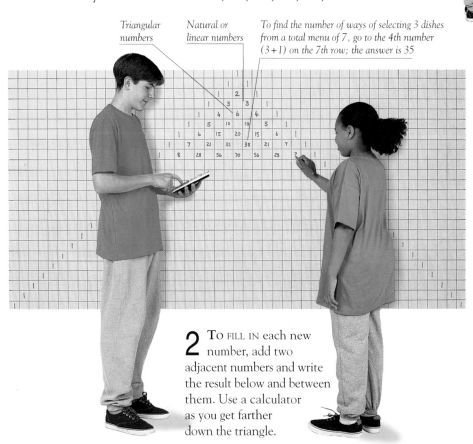

2 TO FILL IN each new number, add two adjacent numbers and write the result below and between them. Use a calculator as you get farther down the triangle.

Patterns in the triangle
When the triangle is completed, shade in each square with an odd number in it and you will see a pattern emerge. This illustration shows the pattern over 30 rows; if you carry on, even more amazing effects occur.

EXPERIMENT
Sissa's reward

Sissa ben Dahir's story shows a demonstration of another number sequence. In the myth, Sissa was a courtier for an Indian king; he invented a game similar to chess. The king offered Sissa a reward for his work. Sissa asked for grains of rice calculated so that one grain of rice was put on the first square of his chessboard, and then the number doubled for each successive square. The king thought Sissa a fool, but little did he know how clever Sissa was! Watch how quickly the numbers grow. We have used chocolate coins in place of rice.

You Will Need
- *chessboard*
- *chocolate coins*

1 PLACE ONE coin on the top left square of the chessboard. This marks the beginning.

Place the single coin on the first square

2 PUT THE COINS ON THE second square. Continue the sequence by doubling the number of coins on each new square. Use your calculator; key in "×2 =" each time. Was Sissa clever in asking for this as a reward? (Answer on p. 186.) You may find that you run out of chocolate coins, in which case write the numbers down until you have reached the last square.

DEMONSTRATION
The Fibonacci sequence

Fibonacci, whose full name was Leonardo of Pisa (1180–1250), was the son of an Italian merchant. On his travels in Europe and North Africa, he developed a passion for numbers. In his major book, *Liber Abaci* (p. 30), he described a puzzle that gave rise to what is known as the Fibonacci sequence of numbers. The sequence is 1, 1, 2, 3, 5, 8, 13, 21 . . . . Each number or term in the sequence is the sum of the previous two ($1 + 1 = 2$, $1 + 2 = 3$, $2 + 3 = 5$, . . .). The Fibonacci sequence can be observed in the arrangement of leaves on a flower or segments in a pineapple or pinecone.

Counting the scales
Look at a pinecone, and you can see that the scales of the cone form regular spirals — some go to the left and some to the right. If you count the numbers of scales at each level, you will find that they follow the Fibonacci sequence.

123 Trick
For this trick, you will need paper, a pencil, and a calculator.
1. On a sheet of paper, make ten rows and ask one friend to write a number on row one.
2. Get another friend to write another number in row two.
3. Now have one friend add the numbers together and write the total in row three.
4. Ask them to continue the series, adding the last two numbers together each time.
5. When your friends get to the seventh row, quickly look at the paper, then multiply this number by 11. Write the number down, and turn it upside down on the table.
6. When your friends have finished adding the 10 numbers, ask them to add all the numbers together.
7. Turn over your answer to show that you already had the answer.
This trick works because in the Fibonacci sequence, the seventh number is one eleventh of the total of the first 10 numbers.

PROPORTIONS

Nature and architecture
*The façade of the Casa d'Oro palazzo, on the Grand Canal
in Venice (left), is aesthetically pleasing not only in its
symmetry (p. 158) but also because of the ratios used in its
design. The spirals of the petals of a daisy-like flower (above)
also conform to a mathematical series named after the
early Italian mathematician Fibonacci (p. 49).*

RATIOS, FRACTIONS, and
percentages express the
mathematical relationships
between the parts of an object
or an idea and the whole. Some
ancient civilizations, most
notably the Egyptians, used
fractions as a crucial part of
their calculations. Nowadays,
proportions, such as those that
govern the physical properties
of gases or electrical circuits, are
an important way of conveying
information in science and
technology, and percentages
are used in economics and
statistics, by people in business,
government, and finance.

COMPARING AMOUNTS

RATIOS ARE USED TO COMPARE various different quantities. They can be employed to specify anything from the ingredients used in preparing a meal to the mass of atoms. Some ratios, such as the Golden Section, have been found to occur throughout mathematics and nature, and the study of these ratios may help us to understand the living world.

The Italian artist Masaccio (1401–28) was one of the first Renaissance painters to depict perspective by drawing near and far lines in different ratios. In his fresco of the Trinity, in Florence, Italy, the tops of the four columns around Christ appear to form a square.

A ratio indicates the relative sizes of two or more quantities — how many times bigger one quantity is than another. The ratio of two quantities can be calculated by dividing them. For example, if a cake contains 6 oz (180 g) of flour and 2 oz (60 g) of butter, it is said to contain flour and butter in a ratio of 3 to 1 (180 ÷ 60 = 3).

Using ratios
Ratios describe many familiar quantities. For example, they are at the heart of money exchanges, which control the relative values of currencies around the world. These values, known as rates of exchange, vary from hour to hour according to the economics and politics of different countries. Large sums of money can be made and lost when traders buy and sell currencies in response to these variations.

Tourists must convert their money to the currency of the country that they visit. The rate of

Ratios in banking
Coins were weighed and compared at currency exchanges in Europe as long ago as the 12th century. All international currency dealings are still based on the changing ratios between currencies, which are known as the exchange rates.

exchange enables them to work out exactly how much something costs in their home currency.

Ratios only compare quantities defined in the same units. Speed, for example, is not a ratio. It is calculated by dividing the distance that an object travels by the time taken, and is given in units such as miles or kilometers per hour (mph or km/h). The symbol "/" or "per" indicates the division between units of distance and units of time.

Ancient Egypt
The ancient Egyptians were some of the earliest people to use ratios. Extracts from the Rhind Papyrus (p. 14) describe how ratios were used over 3,500 years ago to make sure that the Pyramids faced in the right direction. Because of their astrological beliefs, the Egyptians wanted to align their square-based pyramids (p. 153) exactly along the compass points. The north–south line was easy to find using shadows cast by the Sun. But the east–west line could only be found by constructing a line at right angles to the north–south one. So the Egyptians constructed this line using ropes that were divided into sections using knots. To minimize errors, the pyramid

builders used special people called rope fasteners who divided the ropes into sections that had exactly the right ratios.

Written forms
The ancient Egyptians wrote ratios, as we do now, using fractions. A fraction is one whole number written above another: ¾ (three-quarters), ¹⁹⁄₄₀ (nineteen-fortieths), and –⁵⁄₇ (minus five-sevenths) are all examples of ordinary fractions.

Colors in paint
The colors in art materials and printing ink can be mixed by hand or pre-mixed in a given ratio to produce exactly the same shade every time.

Mathematicians call the lower number of any fraction the "denominator" and the upper number the "numerator." The denominator tells us the number of parts into which an object is divided, and the numerator tells us how many parts are used. For example, ³⁄₁₀ of a cake is three parts of a cake that has been divided into ten equal parts.

Unlike us, the Egyptians wrote "unit" fractions, with numerators of 1 only. Other fractions were expressed by adding such fractions together. For example, they would write the fraction ⅖ as ⅓ + ¹⁄₁₅. The one exception was the ratio ⅔, which they wrote directly.

The ancient Chinese worked with fractions around the same time as the Egyptians. They called

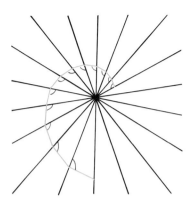

The equiangular spiral
This spiral is formed on equally spaced radii drawn from a central point. The spiral cuts through any radius through the center at the same angle each time.

their denominators and numerators "mothers" and "sons." The Romans and Babylonians developed fractions that had the same denominator but could have any number as the numerator. Roman fractions had 12 as the denominator, and fractions in the Babylonian system had a denominator of 60. Using these systems, certain fractions could only be given approximately: ⅟₇, for example, is between ¹⁄₁₂ and ²⁄₁₂ in the Roman system.

Today, ratios can sometimes be written using the symbol ":", as in 6:11 (a ratio of 6 to 11). The ":" symbol, first used by the British mathematician William Oughtred (1575–1660) in 1631, has a different meaning from the dividing line in fractions. The fraction ¾ refers to three parts of a whole divided into four parts. In contrast, 3:4 means three parts and four parts of a whole divided into seven parts. The ":" symbol can be used in ratios of more than two quantities. Three gears on a bicycle, for example, may have numbers of teeth in the ratio 1:3:5.

Simplifying

Ratios can be simplified by writing them using smaller numbers. For example, the ratio 770:990 can be written more simply as 7:9. Alternatively, many different

ratios can be written using a common number. When ratios are expressed using the number 100, they are called "percentages." These are written using the sign "%". For example, because ³⁄₁₀ is equal to ³⁰⁄₁₀₀, it can also be written as 30 percent, or 30%.

Percentages allow us to compare ratios quickly. For example, it is difficult to see instantly whether ⁶⁄₁₅ is bigger or smaller than ⁴²⁄₁₂₀. However, when they are both expressed as percentages, 40% and 35%, the task becomes much easier.

Ratios in nature

The Greek mathematician Pythagoras founded a school of philosophers (p. 60). They studied numbers called "**ratio**nal numbers" — numbers that are given as the ratio of two numbers.

They believed that numbers could explain the nature of the Universe, and this idea led them to investigate ratios in the physical world. One of their most important investigations concerned vibrating strings. They found that strings that had lengths in the ratio 1:2:3:4 could be used to make all the known "consonant intervals" (pairs of notes that are pleasing to the ear). This discovery supported the Pythagoreans' belief in the importance of numbers.

The Pythagoreans also studied natural ratios using geometry. One number that they could not define exactly was the ratio of the circumference to the diameter of

a circle (p. 134). We now call this ratio "pi" (π). Numbers such as pi are called "ir**ratio**nal" numbers (p. 38) because they cannot be expressed perfectly as a ratio. The discovery of irrational numbers so shocked the Pythagorean scholars that there was uproar and even threatened assassination of the mathematician who first described the concept.

Another ratio that interested them was the ratio of the Golden Section (p. 58). The ratio, which is approximately 1:0.61803, occurs in nature, and has been used for thousands of years.

Shadow ratio
By finding a small plant and measuring the ratio of its shadow length to its height in certain light conditions, you can, under the same light conditions, work out the height of a much taller plant that could not otherwise be measured (p. 54).

Investigators of this ratio found that buildings and paintings seem most pleasing to the eye when they are designed to the proportions of the Golden Section. Similarly, music seems satisfying when the time spans of its different sections are proportioned in this way. It occurs in natural forms such as the spirals on sea shells and the rows of petals in flowers. Scientists are still fascinated by this ratio. At the end of the 20th century, many expect to find it within the patterns encoded in DNA (p. 149), the basis for all living things.

A natural ratio
The shells of mollusks such as the nautilus have spirals that naturally follow an equiangular spiral in the Golden Section. The illustration top left mimics the shape of the shell, showing how mathematics is also found in nature.

Comparisons

A RATIO IS A COMPARISON of different amounts of the same thing. For example, for a soft drink in which the ratio of syrup to water might be one part to four parts, the volume might be written as 1:4. Sometimes a ratio is written as one number over another, such as ⅖. Ratios are widely used in science and economics. Examples include a business calculation called the "price to earnings ratio" (P/E). Some ratios are vital to mathematics. For instance, the ratio between the circumference of a circle and its diameter is the same for all circles, and can be given approximately as 3.142:1 or exactly as π:1 (p. 134). Other important ratios are the sine, cosine, and tangent in trigonometry (p.127).

EXPERIMENT
Reducing air pressure

Many common scientific phenomena have been reduced to simple proportions. One such ratio is a law governing the relationship between the pressure and volume of a gas (such as air) if the temperature is constant. In this experiment, air is pumped out of a bottle in order to decrease the pressure inside. The marshmallows inside the bottle contain trapped bubbles of air. Watch what happens to them as the air expands to fill the space.

YOU WILL NEED
● *vacuum pump and vacuum seal for wine bottles*
● *marshmallows*
● *clear bottle*

1 FILL the bottle three-quarters full with marshmallows. You may have to cut the marshmallows in half lengthwise so that they fit through the neck.

2 INSERT the seal and attach the pump. Pump out the air. The marshmallows will swell, showing how the volume of the air inside them increases in proportion to the decrease in air pressure.

3 CONTINUE to pump until the marshmallows fill the bottle, then open the valve to let the air out. See the marshmallows shrivel as the air pressure increases again and the volume of air in them decreases.

EXPERIMENT
Ratios in shadow

The ratio of the height of an object to the length of its shadow will always be constant as long as the angle of the light on it is constant. Using this knowledge, it is possible to calculate the height of a plant (or shrub) that is too tall to be measured physically by knowing only the length of its shadow.

YOU WILL NEED
● *pen* ● *notepad*
● *tape measure*
● *potted shrub*

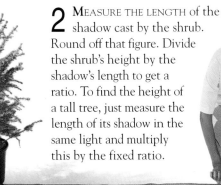

1 MEASURE THE HEIGHT of the shrub, including the pot, with your tape measure. Round off the figure up or down to the nearest 4 in (10 cm).

2 MEASURE THE LENGTH of the shadow cast by the shrub. Round off that figure. Divide the shrub's height by the shadow's length to get a ratio. To find the height of a tall tree, just measure the length of its shadow in the same light and multiply this by the fixed ratio.

EXPERIMENT
Population density

In science, density is defined as the mass of a given object divided by its volume (p. 103). You can use the same principle to find population density, which defines the average number of people living on a given area of land. It is used to describe how crowded a country or city is. This example shows a group of countries in Southeast Asia, from Mongolia to Hong Kong. You could study these nations, or you could look at countries or towns in your area to see if you live in a crowded or a sparsely populated place.

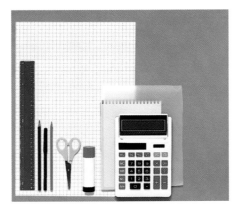

You Will Need
- ● ruler ● pencils ● pen ● scissors ● glue
- ● calculator ● notepad ● tracing paper
- ● graph paper ● sheet of poster board

Exchanging money

In international business deals, or when people travel abroad, money needs to be changed from one currency to another. The exchange rate (the ratio of one currency to another) is controlled at currency exchanges such as the one shown above. Travelers use this ratio to compare prices abroad to the costs of equivalent items at home.

1 Using an atlas, trace the shapes of the countries to be compared. Make a note of each country's population and its area. The areas for all the countries should be in the same units: either square miles or square kilometers.

2 Color the shapes of the countries, then cut them out. (Write the name on the back of each shape as a reminder.) Calculate the population density of each country by dividing its population by the land area.

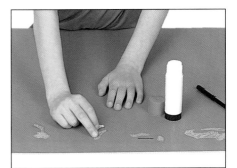

3 Glue the country shapes in a line at the top of the sheet of poster board, spacing them well apart so that you can see clearly which is which. Write the name of each country just above or beside its shape.

4 Make one square of graph paper equal to one person per square mile (square kilometer). On the graph paper, count the squares that represent each country's population density, and cut out the blocks or strips.

This big piece of graph paper represents the population density of just one country in Southeast Asia

5 Glue each strip or block below the appropriate country. If you have studied Southeast Asian countries, can you see which one has the highest population density? (Answer on p. 186.)

Using ratios

RATIOS ARE OFTEN USED for convenience when the mathematical quantities themselves are awkward to handle. For instance, it is easier for astronomers to classify stars according to apparent magnitude (see right), based on ratios, because some stars may be billions of times brighter than others. The Richter scale is another example: the magnitude of an earthquake (between 1 and 10) is based on the energy of the quake, and each magnitude is 10 times more powerful than the previous one. Ratios also play an important part in number theory (p. 38). Rational numbers are those that can be expressed as the ratio of two whole numbers, such as ½. In contrast, irrational numbers (p. 38), such as $\sqrt{2}$, cannot be exactly expressed in this way. When absolute quantities are not essential, ratios can again be of use. In cookery, for example, taste and texture are affected by the relationship between ingredients rather than the absolute amounts used.

Apparent magnitude

Sirius, in the constellation of Canis Major, is the brightest star in the night sky. Stars are classified by apparent magnitude — their brightness as viewed from Earth. This system is based on the ratio 1:2.512. A star of magnitude 1 is 2.512 times as bright as one of magnitude 2. Stars of magnitude 1 are 100 times as bright as those of magnitude 6. Some are so bright that they have a negative figure. Sirius has a magnitude of –1.6, and the Sun has a magnitude of –26.7; the Sun appears 10 billion times as bright as Sirius.

EXPERIMENT

Mixing and matching tints

Making colors and matching them is so important in art and industry that scientists have developed charts of standard colors to use as references. You can create your own tints, using different ratios of the primary colors yellow, red, and blue, and compile a reference chart showing the ratios used to produce each color. You will then be able to reproduce those particular tints exactly at a later date.

YOU WILL NEED
- water ● powder paints
- paint containers
- palette ● paintbrush
- felt-tip pens
- spoon ● paper

1 USING THE SPOON to measure exact quantities of paint, put two parts blue and one part yellow in one container to make green. Mix this with a little water. Then, in a new container, make up another shade of green using one part blue and two parts yellow.

2 MAKE UP A REFERENCE chart so that you can mix identical colors another time. Use the pens to show the ratios, coloring in one square per spoonful of paint in each primary color. Paint a small brushstroke of the mixed color beneath the appropriate ratio.

International standards in color
This man is using one of the internationally recognized standard color charts to match a specific color of paint. The exact color can then be ordered by quoting the number on the chart. In the color charts, the amounts of each color used to make a particular shade are given as percentages of one of the three primaries.

EXPERIMENT
A recipe for shortbread cookies

Some recipes use a particular ratio of ingredients so that the amount or weight of food being cooked can be varied without the taste or texture changing. You can make as many shortbread cookies as you like by altering the container that you use as your measure but keeping the proportion 3:2:1.

YOU WILL NEED
- *mixing bowl* • *3 measures of flour*
- *2 measures of butter or margarine*
- *1 measure of sugar* • *wooden board*
- *cookie cutter* • *knife* • *nonstick baking sheet*

1 MIX ALL the ingredients in the bowl. Use the knife to cut up the butter, then knead with your fingers until the flour and sugar are blended into the butter and the mixture is sticky.

2 TIP THE MIXTURE out in front of you. Flatten it to about ½ in (1 cm) thick. You may need to put flour on the work surface to keep the mixture from sticking. Use the cutter to make cookies, and lay them on the baking sheet. Make sure their edges do not touch.

3 BAKE THE COOKIES at 250°F/120°C for about 20 minutes or until light brown. Use a cloth or oven mitt to take the baking sheet out of the oven, and place it on a wooden board so that the tray does not burn the work surface.

EXPERIMENT
Making music with air and water

Ratios are widely used in music. For example, in every key there are the same set ratios between the pitches of notes in an octave (p. 60). You can produce a range of notes when you blow across the tops of a series of bottles, by filling the bottles with different ratios of liquid to air. The note will be lower the more air there is in the bottle.

YOU WILL NEED
- *ruler* • *drink stirrer*
- *jug of water* • *food coloring* • *adhesive tape*
- *scissors* • *5 identical straight-sided bottles*

1 MEASURE THE DEPTH of a bottle from the base to the start of the neck. Divide this measurement into fifths. On one bottle, use tape to mark ⅕. Place pieces of tape at the appropriate levels on the other bottles to show ⅖, ⅗, ⅘, and 1.

2 FILL EACH bottle with water up to the tape mark. If you wish, you can add different shades of food coloring to make the water levels easier to see. Take care when handling the food coloring, because it may stain your skin or clothes.

3 YOU CAN NOW PLAY a tune on the water-filled bottles by blowing across the top of each one in turn. Listen to the different sounds that result from the various combinations of air and water.

Divine proportions

THE GOLDEN SECTION is a special ratio that is approximately equal to 1.618:1, or $\frac{\sqrt{5}+1}{2}$:1. It has been used in art and architecture for centuries, and is also found in nature. The Golden Section divides a line at a point so that the ratio (p. 54) of the smaller part of the line to the greater part is the same as the ratio of the greater part to the whole line. Also known as the Golden Mean, this proportion is said to be the most pleasing to the eye. The Greeks were intrigued by this special mathematical relationship. Before that, the Egyptians certainly had a "sacred ratio;" on the Great Pyramid at Gizeh in Egypt, the ratio in the height of a face to half of a base is 1.618:1.

▥ Modern methods

Many architects have used the Golden Section in the designs of their buildings. The United Nations Building, built in New York in 1952, is a modern example of architecture in which the Golden Section has been applied. The ratio of the height of the building to the length of its base is 1.618:1. The French architect Le Corbusier (p. 118) even used the Golden Section in his domestic designs.

EXPERIMENT
Investigating the Parthenon

The Parthenon temple, part of the Acropolis in Athens, Greece, is one of the most splendid classical buildings, and is considered by many to be almost perfect. It was designed by the ancient Greeks as a shelter for the goddess Athena. It has lost its entablature (the triangular roof section), but if that is included, does the building conform to the Golden Section? If it does then this would show that as long ago as the 5th century B.C., architects were aware of the aesthetic qualities of the Golden Section. We do know that Renaissance artists used this special ratio, calling it the Divine Proportion.

YOU WILL NEED
● ruler
● calculator
● pens
● set square
● picture of Parthenon

Parthenon engraving
This engraving shows what the Parthenon used to look like. Photocopy it, enlarging it by about 50 percent. Measure the photocopy so that you can find out if it conforms to the Golden Section.

1 PHOTOCOPY the engraving of the Parthenon. On the photocopy, rule a rectangle around the outer limits of the building, including the entablature (the triangle at the top).

3 MARK THE height measurement on the rectangle's longer sides, then draw a line on the shape to complete the square. Use the set square to check that this is a true square.

2 MEASURE the height of the building, and write the measurement beside one of the rectangle's shorter lines. Let the height measurement equal one side of a square.

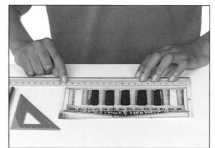

4 MEASURE the line at the base of the building. Using a calculator, divide the length of one side of the square by this figure to see if the ratio conforms to the Golden Section.

EXPERIMENT
A natural spiral

The Golden Section is not represented only in rectangular shapes, such as buildings. It can also be used to form a beautiful spiral (pp. 146–147). This special spiral is often found in nature — in the whorls of a sunflower, the arrangements of leaves on branches, or the scales on pinecones (p. 49). The spiral is based on adjacent squares drawn inside rectangles with sides in the Golden Section. To draw your own spiral, build up increasingly large squares on graph paper, then draw a quarter of a circle in each square. The finished spiral mirrors those found in nature.

YOU WILL NEED
- *ruler* ● *pen* ● *pencils*
- *pair of compasses*
- *graph paper*

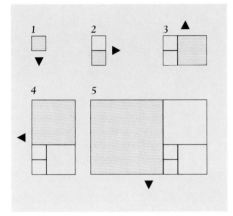

The template
Start with one square on your graph paper. Follow the arrows to add new squares on one side of the previous squares. Repeat this step several times.

1 COLOR the squares so that they stand out clearly. Use your compasses to draw a quarter-circle in each square.

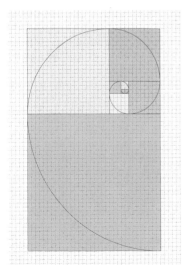

2 JOIN THE curved lines smoothly to form the spiral. All the rectangles in the pattern have the same proportions.

Spiral on a nautilus shell
The nautilus, a sea creature, is a mollusk, like the snail. Its shell illustrates how nature follows the rules of proportion. Compare your drawn spiral to the one in this shell; it should closely mirror the natural spiral.

EXPERIMENT
Making the Golden Section

Here is a simple way to create the ratio. By making a regular pentagon, which has five equal sides and equal internal angles (p. 128), you can investigate how the diagonals divide each other in the Golden Section. This experiment also shows how to make a pentagon just by tying a simple knot in a piece of paper.

YOU WILL NEED
- *scissors* ● *pencil*
- *ruler* ● *15-in*
 (38-cm) strip of paper

1 FOLD the strip of paper into a single knot. Pull the ends gently as far as they will go.

2 PLACE THE KNOT on the table and gently flatten it with the palm of your hand. Sharpen the edges with your fingernail, then cut off the flaps to make the pentagon shape.

3 DRAW TWO CROSSING DIAGONALS between points of the shape. Measure the longer part of one, and divide this number by the length of the shorter part. The ratio will be about 1.618:1.

What is a fraction?

A FRACTION IS WRITTEN by putting one number over another and dividing by a line, as in ⅓. The top number is the numerator; it shows how many parts are in the fraction. The bottom number is the denominator and shows the number of parts into which the whole is divided. Fractions can be less than 1, such as ⁵⁄₇, or greater than 1, such as ⁹⁄₇ (which can also be shown as 1²⁄₇). The fraction ⁵⁄₇ is called a "proper" fraction, because the numerator is less than the denominator; ⁹⁄₇ is known as an "improper" fraction. The number 1²⁄₇, a whole number with a fraction, is a "mixed" number. Fractions are usually expressed in their lowest terms, when both the numerator and the denominator have been divided by the highest common factor (p. 183). For example, ¹²⁄₃₆ is usually written as ⅓, with both numbers divided by their highest common factor (12).

Music and the Pythagoreans

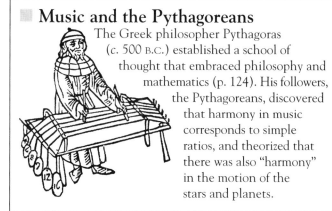

The Greek philosopher Pythagoras (c. 500 B.C.) established a school of thought that embraced philosophy and mathematics (p. 124). His followers, the Pythagoreans, discovered that harmony in music corresponds to simple ratios, and theorized that there was also "harmony" in the motion of the stars and planets.

EXPERIMENT

Musical instrument

On stringed instruments, the ratios between notes can be expressed as fractions of the string lengths, as shown on this simple sound box.

Adult help is advised for this experiment

YOU WILL NEED
- *craft knife* ● *pair of compasses*
- *1 large pencil, 2 small pencils*
- *scissors* ● *adhesive tape* ● *long rubber band* ● *brass fasteners* ● *long shoe box or flower box*

1 WITH A PAIR of compasses, or using a round object as a template, draw a large circle near one end of the box. Ask an adult to cut out the circular shape using a craft knife. Use tape to attach a small pencil at either end of the box, 1½ in (4 cm) from each edge.

2 INSERT THE TWO brass fasteners in the underside at either end. Twist one part of the rubber band around one fastener, then stretch it over the pencil to the other end of the box. Secure the rubber band around the other fastener so that the band is taut across the box.

3 PLUCK THE rubber band over the sound hole of your musical instrument. This makes the air vibrate inside the box, which amplifies the original sound. On other musical instruments, strings can also be bowed (as in a violin) or struck (as in a piano). Listen to the note.

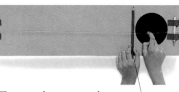

4 TUCK A long pencil under the elastic band about a quarter of the way down the box, at the end near the sound hole. Pluck the elastic band again, and listen to the note produced. You should notice a difference.

Hold the pencil firmly

5 MOVE THE pencil halfway down the box, and pluck the band. You will hear that the note has dropped by an octave.

The length of the vibrating part is doubled to make the note drop by an octave

EXPERIMENT
Making a photometer

A photometer is an instrument used to compare brightness. You can make your own as a guide to the relative brightness of two lamps. Light spreads over an increasingly large area as it travels, and the illumination reduces with the square of the distance (p. 40). You can find the approximate brightness of one lamp if you know the brightness of the other, by adjusting the frame and using this ratio:

$$\frac{\text{Brightness of light 1}}{\text{Brightness of light 2}} = \frac{(\text{distance to light 1})^2}{(\text{distance to light 2})^2}$$

YOU WILL NEED
- *2 lamps with bulbs of different wattages*
- *butter* • *adhesive tape* • *scissors*
- *craft knife*
- *pencil* • *paper*
- *poster board*
- *ruler* • *calculator*
- *notepad*

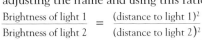

Jazz and maths

Gottfried Leibniz (p. 47) wrote: "Music is a secret arithmetical exercise and the person who indulges in it does not realize that he is manipulating numbers." Jazz may sound abstract, but two formal mathematical patterns dominate. One comprises four eight-measure sections; the other, with roots in African-American folk music, is the 12-bar blues form. Jazz musicians such as this clarinet player improvise within their chosen style, working on the principle that a variety of melodies can fit the chords of a tune.

To the viewer the spot appears bright

From the front the same spot appears dark

1 MAKE A FRAME from poster board (p. 163) and tape the paper to it. Dab a spot of butter on the paper and shine a lamp behind it. The spot will look bright because it allows light to pass through.

2 SHINE THE LIGHT on the front of the paper. The fat spot will appear darker because the fat will let some light through instead of reflecting it.

3 TURN BOTH LIGHTS on. Put the paper between them. Move the frame until the fat spot seems to vanish. This happens when the spot receives equally intense light from both lamps. Measure the distances between the spot and each bulb. Work out the squares of these distances and show them as a fraction, using the formula given above. This fraction is equal to the relative brightness of the two bulbs.

Finding fractions

FRACTIONS OCCUR COMMONLY throughout mathematics, science, and the natural world. They have been in use since the time of the ancient Egyptians (p. 52). In A.D. 1202 the mathematician Fibonacci (p. 30) described a complicated system of fractions for use in working out currency exchange. He also created tables of conversion from common fractions, such as ⅜, to unit fractions where the numerator is always 1, such as ⅛. Decimal fractions, such as tenths, hundredths, and thousandths, came into common usage only after Simon Stevin (1548–1620), a mathematician from Bruges, Belgium, published his work *De Thiende* (The Tenth) in 1585, and the decimal point (p. 34) came later. Since that time the use of decimals has become widespread in mathematics and in everyday life, while calculation with common fractions has greatly diminished.

The hidden iceberg

Icebergs, such as this one in Antarctica, are formed when parts of a glacier break off as the glacier reaches the open sea. Icebergs can tower up to 492 ft (150 m) high above the water, yet almost nine-tenths of their mass and three-quarters of their height remain hidden below the surface. For this reason icebergs are renowned for their potential danger, despite their imposing and beautiful appearance. They are a serious hazard to ships, which can crash into them; the *Titanic* (p. 81) was the largest and best-known ship to sink after colliding with an iceberg.

EXPERIMENT
Predicting what ice will do

The Archimedes principle is a physical law that makes it possible to predict if a body will float. When an object floats in water, the force supporting the object is called the upthrust. The Archimedes principle states that the upthrust on an object is equal to the weight of fluid displaced by the immersed part of the object. In this experiment, ice cubes are dropped into a glass of water. When water freezes, it expands in volume by a certain fraction (opposite page). Not all of each ice cube is under water, so does this mean that the water level will rise when the ice melts?

YOU WILL NEED
● *glass of water* ● *ice cubes*
● *scissors* ● *adhesive tape*

Puzzle

Fractions are usually reduced to their lowest terms, but in this puzzle you can reverse the process and find several different ways to write a fraction, say, equivalent to ½. Try to write a fraction that is equal to ½, using all the digits — 1, 2, 3, 4, 5, 6, 7, 8, and 9. (Answers on p. 186.) One example is ⁶⁷²⁹⁄₁₃₄₅₈. There are at least six possible answers. Try the same game again, this time using single digits to find other fractions, such as the equivalents of ⅓.

Checking the level
Put some water in the glass and add three ice cubes. Mark the water level on the side of the glass with a piece of tape. Put the glass to one side and let the ice melt. Check the water level after the ice has melted. Has it risen or fallen, or has it stayed the same? (Answer on p. 186.)

EXPERIMENT
Freezing water

One important property of water is that it expands when it freezes: the volume increases, lowering the density, which is why ice floats in water (p. 103). This experiment will help you calculate the fractional increase in volume when water freezes. Measure the increase between the first water level and the second (frozen) level, then find the fractional increase using the formula:

YOU WILL NEED
● *ruler*
● *plastic bottle*
● *jug of colored*
  water ● *pen*
● *adhesive tape*

$$\text{Fractional increase in volume} = \frac{(\text{2nd reading} - \text{1st reading})}{\text{1st reading}}$$

1 STICK A LONG PIECE of tape to the side of the bottle. Starting at the base, measure the side with a ruler and make marks with a pen up the tape at ½-in (1-cm) intervals.

2 POUR WATER into the bottle up to the 3 in (or 6 cm) line. Put the bottle upright in a freezer, until the water freezes. Which mark does the frozen water reach? Work out the fractional change using the formula on the left. (Answer on p. 186.)

▦ Buying for the family

Fractions are used in stores to give convenient quantities for us to buy. For instance, we may talk about half a pound or a quarter of a kilo of cheese. But goods are also sold in different amounts in containers of varying sizes. To find out the cost of each unit of volume or weight in a container and to see if it is cheaper to buy larger or smaller sizes, use the following fraction. The lower the price per unit, the bigger your savings are.

$$\text{Price per unit} = \frac{\text{Price of container}}{\text{Number of units in container}}$$

▦ How much land is there?

A proportion described as a fraction is often easier to use than one shown as a decimal because fractions are more easily visualized. The fraction of the Earth covered by land is surprising. To calculate this fraction, you will need an atlas or other reference book to give you the data for the area of each continent and the area of the Earth's surface.

You can try to estimate the fraction by looking at a globe

Working out land area
Write down the surface area of the Earth. Now add up the areas of the seven continents (including Antarctica) to find the approximate total land area on the Earth. Both areas should be given in the same units, either square miles or square kilometers. Write the land area as a fraction of the Earth's surface:

$$\frac{\text{Land area}}{\text{Total surface area}}$$

Round off both figures to the nearest hundred, then reduce the fraction to its simplest terms; this gives an estimate of the fraction of the Earth covered by land. Check your answer on p. 186.

By the hundred

A PERCENTAGE IS A WAY of representing a fraction as part of 100. Twenty percent, for example, is equal to $^{20}/_{100}$, and is written as 20%. To calculate, say, 23 units as a percentage of 50 units, the fraction is shown as part of the whole and multiplied by 100%. For instance, $^{23}/_{50} \times ^{100}/_{1}\% = ^{23}/_{1} \times ^{2}/_{1}\%$, or 46%. Percentages are used in school exams to calculate how well a student has done; in finance to calculate profits and losses; and by governments to determine taxes. They may be used to compare similar types of figures, but the comparison can be deceptive. If nurses, say, have a 5% pay increase, while civil servants have 2%, the nurses may seem to be better off; but if the increases are 5% of a salary of 100 (=5) and 2% of 300 (=6), then the civil servants gain more.

INVESTIGATING NUMBERS
Using percentages

Experts often use percentages to confuse an issue, perhaps by quoting figures without saying how they were derived. An example is the statement: "prices rose by 50%, to be reduced later by only 33⅓%." At first glance, it seems that the price is still higher than it was originally.

1 The original price is $100. In year 1, the price rises by 50%.

$50\% \times \$100$
$= ^{50}/_{100} \times \$100$
$= \$50$

2 This rise produces a new price.

$\$100 + \$50 = \$150$

3 In year 2, this price is reduced by 33⅓%, or ⅓.

$33\tfrac{1}{3}\% \times \$150$
$= \tfrac{1}{3} \times \$150$
$= \$50$

4 The reduction brings the price back to $100. It is important to know the figure from which a percentage is originally calculated.

$\$150 - \$50 = \$100$

EXPERIMENT
Color wheels

Painters often mix the colors yellow, red, and blue to make up other colors. Professional printers (p. 56) use the primary pigments magenta, cyan, and yellow (along with black) to make full-color images. This experiment shows how to mix colors without mixing paints, by painting them separately on a wheel and then spinning the wheel very fast, so that the colors appear to blur together. Draw another wheel, and change the percentage painted in one color; see if the colors produce a different effect.

YOU WILL NEED
● *ruler* ● *pencil* ● *pair of compasses* ● *scissors* ● *protractor* ● *string* ● *poster board* ● *red, blue, yellow paints* ● *paintbrush*

1 DRAW A large circle on the poster board, setting the pair of compasses at a radius of at least 4 in (10 cm). Cut out the circle.

2 USE A protractor to divide the circle into thirds — each part being 33⅓% of 360° (120°). Draw radii to mark the edges of the thirds.

3 PAINT each third of the circle in a different primary color, following the edges as carefully as possible. Leave the paint to dry.

Watch the wheel closely as it spins

4 MAKE two holes on either side of the center and about ½ in (1 cm) apart. Push a piece of string about 48 in (120 cm) long through both holes and knot the ends.

5 HOLD THE ENDS of the loop. Wind up the string clockwise on one side of the wheel and counterclockwise on the other. Let the wheel spin, and watch the effect.

EXPERIMENT
Ready reckoner

This ready reckoner is shaped like a wheel. Each spoke in the wheel has windows that show the different ways of calculating a particular percentage — mentally and on paper, as a fraction, and as a decimal.

 Adult help is advised for this experiment

YOU WILL NEED
● *craft knife* ● *ruler* ● *adhesive tape* ● *bradawl* ● *scissors* ● *brass fastener* ● *pencil* ● *pen* ● *2 pieces of poster board* ● *cutting mat*

1 **P**HOTOCOPY the templates (right), enlarging each by 48 percent. Tape the copy of the front wheel template to a piece of poster board. Cut it out.

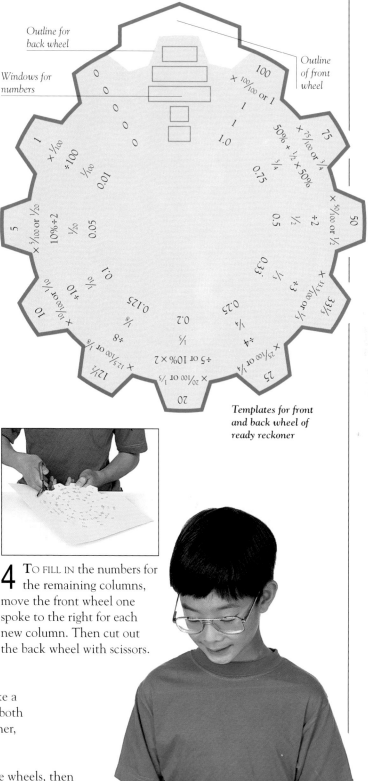

Outline for back wheel

Windows for numbers

Outline of front wheel

Templates for front and back wheel of ready reckoner

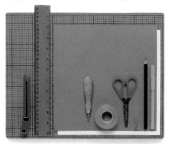

2 **P**OSITION this template on the front of the wheel again. Ask an adult to cut out the five windows with a craft knife, slicing through both the paper and the poster board.

3 **C**UT OUT THE back wheel template. Lay it on the other piece of poster board, and draw around it. Align the two wheels and copy the 0's from the template into the windows.

4 **T**O FILL IN the numbers for the remaining columns, move the front wheel one spoke to the right for each new column. Then cut out the back wheel with scissors.

5 **A**SK AN ADULT to make a hole at the middle of both wheels of the ready reckoner, using a sharp implement.

6 **A**LIGN the wheels, then push the brass fastener through both central holes to join them. On the front, write "%" by the top window, "written" by the next one down, then "mental," "fraction," and "decimal" (in that order) by the other windows. The windows show which calculation is involved for each percentage.

ALGEBRA

Symbols and formulas

In writing systems such as the ancient Greek "Linear B" (above), symbols on clay tablets were used to keep numerical records. Today architects and engineers apply complex algebraic equations when they design a building (left), both to plan the structure and to allow for the stresses that the building will experience.

ALGEBRA IS A FORM of mathematics that simplifies difficult problems by using letters to represent numbers, many of them unknown, in calculations. It is also the basis for the mathematical expression of many scientific formulas. Unknowns such as the amount of concrete needed for a building, the weight of grass seed for a lawn of a given area, or the number of loaves of bread required for a banquet can all be found using algebra.

HISTORY OF ALGEBRA

WHEN MATHEMATICIANS need to find a solution to a problem, they often use a letter to represent the number they are seeking. This technique, called algebra, enables many different problems in the real world to be solved. Today, algebra is necessary not only in mathematics but also in science, economics, finance, and technology.

Algebra enables mathematicians to investigate problems that can be shown in a mathematical form. An unknown number in a problem is represented by a letter, usually *x*. Such letters are called "variables" because they can stand for any unknown number. Variables can be arranged to form an "algebraic equation" (p. 72), so called because it uses algebra to show things that are equal to each other. One example of an equation is the expression $x = 3x - 2$. Once a problem has

Rhind Papyrus
The Rhind Papyrus (p. 14), the largest mathematical text left from ancient Egypt, shows an unknown number referred to as a "heap." The technique shown in this papyrus was not the one used today, but nevertheless it is a type of algebra.

been written in algebraic form, the variables in it can be treated in a similar way to numbers. Because both sides of an algebraic equation are equal, they remain equal if they are both changed in the same way. The number that can be substituted for *x* to make both sides equal is called the solution to the equation.

Early algebra

Algebra has evolved since the time of the ancient Egyptians more than 3,500 years ago. Examples of it can be seen on the Rhind Papyrus (p. 14). The Egyptians wrote out problems in words, using the word "heap" to represent any unknown number.

Around 300 B.C., the ancient Greek scholar Euclid (p. 114) wrote his books *Elements*; in them he included several "identities" (algebraic equations equal for all numbers) that he developed by studying geometric shapes.

The ancient Greeks wrote problems out in full if they could not solve them using geometry. This method, called "rhetorical algebra," limited their ability to solve detailed problems. In the 3rd century A.D., Diophantus of Alexandria (*c.* A.D. 250) wrote a book called *Arithmetica*, in which he used symbols for unknown numbers and for operations such as addition and subtraction.

The influence of the Muslims
By the 8th century, the Islamic empire stretched from India to Spain, and included cities such as Cairo. The Muslims introduced Greek and Hindu mathematics to Europe.

His system was not fully symbolic, it was somewhere between Euclid's system and the one used today. For this reason, it is known as "syncopated algebra."

The Arab influence

Although they closed the last of the ancient Greek schools of learning when they invaded Alexandria in A.D. 641, the Arabs preserved and developed Greek mathematical ideas for many centuries. They brought the Greeks' ideas to Western Europe after settling in Spain in A.D. 747.

The Arabs first encountered these ideas when they met Greek doctors working in Arab cities. They also became familiar with the work of Hindu scholars in India. Two of the foremost scholars were Brahmagupta (598–660) and Arya-Bhata (*c.* 475–*c.* 550). Among other discoveries, Brahmagupta, an astronomer, devised many identities for the areas and volumes of solids. Arya-Bhata created tables of sines (special ratios) and developed a form of syncopated algebra like the system devised by Diophantus.

Once Arab scholars had access to Greek and Hindu ideas, they began to develop new techniques of their own. The most significant contribution to algebra was made by Muhammad al-Khwarizmi (*c.* 780–*c.* 850). In about A.D. 830, he wrote three books on mathematics. The most important

François Viète (1540–1603)
A French lawyer, Viète studied mathematics only in his leisure time. He devised a new system of algebra and a formula for finding pi (p. 134).

was titled *Hisab al-jabr wa'l muqabalah* (Calculation by Restoration and Reduction). "Restoration" means simplifying an equation by performing the same operation on both sides. "Reduction" involves combining

Chess and algebra
A game as logical as chess can be broken down into simple procedures; moves can be analyzed using coordinates (p. 75).

different parts of an equation to make it simpler. These are both essential techniques in algebra today. In fact, al-Khwarizmi's ideas have been so influential that the word "algebra" (*al-jabr*) has been derived from the title of his book.

Letters as symbols

During the Renaissance, algebra became popular among German mathematicians. It was not until the 16th century, however, that another system was invented to replace rhetorical and syncopated algebra. In 1591, the French mathematician François Viète created an entirely symbolic system of algebra. In his book *In Artem Analyticam Isagoge* (Introduction to the Analytical Arts) he suggested that consonants (B, C, D, F, and so on) could represent unknown numbers, and vowels (A, E, I, O, U) could stand for numbers that are known.

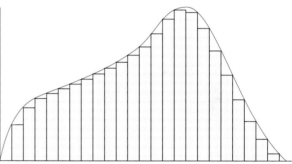

Évariste Galois (1811–32)
Galois gave his name to an important theory on the general solvability of equations. He scribbled down his many mathematical discoveries the night before he died in a duel, aged only 20.

In 1637, René Descartes (p. 74) showed how geometric structures could be converted into algebraic equations. In his book *Discours de la Méthode* (Discourse on Method), he introduced the letters *x*, *y*, and *z* to represent variables, as well as the symbols + and – for addition and subtraction. Descartes' work made it possible to convert the algebra of Euclid and other Greek scholars into a form we can understand and use today.

New solutions

Viète and Descartes produced a highly flexible system of algebra that could be used to solve many problems. Mathematicians and scientists started to employ this system for analyzing aspects of the physical world. At the beginning of the 17th century, many of them turned their attention to analyzing physical quantities that are constantly changing.

As an example, the average speed of an object in motion, such as a ball falling to Earth, can be calculated by measuring how far it moves and dividing this number by the time taken to fall. However, very few objects move at a totally steady speed. For example, it is now known that a ball falling to Earth is constantly accelerating because of the force of gravity. Therefore, scientists often need to specify how fast an object is moving at any instant of time. Newton (p. 71) devised a form of algebra that could be applied to study constantly changing systems. He called this form of algebra "fluxions." Newton corresponded with Gottfried Leibniz (p. 47), who had devised

a similar form of algebra, which he called calculus. Leibniz helped Newton to improve the notation of his fluxions, but historians still debate which of these men was the first to discover an algebra of changing systems. From the supply and demand of goods to the trajectories of rockets, this system of algebra, now known as calculus, has helped people to understand many systems in constant change.

Modern techniques

Mathematicians have also invented new forms of algebra used in very different ways from calculus. One of the most

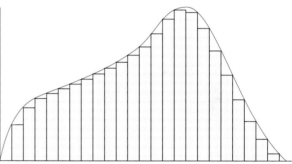

significant new techniques was established by the British mathematician George Boole (1815–64) in his work *Investigation of the Laws of Thought*. His system of algebra, known as Boolean algebra, can be used to write complex logical problems using sets of symbols. Today, computers convert different tasks into a series of simple logical operations using Boolean algebra.

This relatively recent Boolean algebra system, which looks different from other forms of algebra, has so far been applied to topology (p. 164) and probability theory (p. 80), as well as to computing technology.

Principia Mathematica
This major work (p. 71) contained ideas that Newton developed using his "fluxions," a form of algebra that revolutionized science as well as mathematics.

Area under a curve
The mathematics of calculus can be applied to find the area under a curve and to find the rate of change of the curve's slope.

Using letters

ALGEBRA IS IMPORTANT in all areas of mathematics, particularly in calculus. In algebra, constant but unknown quantities or variable quantities are represented by letters. Algebra uses the same arithmetical operations as ordinary math (p. 16); for instance, $a + a = 2a$; $5b - 4b = 1b = b$; $a \times a = a^2$. Many special numbers, such as perfect numbers (p. 44), can be defined in algebraic terms, so that they may be analyzed more thoroughly. Variables can often be put into an equation to show the mathematical relationship between them. For example, the area of a rectangle (A) is expressed as $A = b \times h$, where b is breadth and h is height. Equations enable scientists to predict results such as the current in an electric circuit at a given voltage, or the upthrust, or lift, acting on the wings of an airplane at a certain speed. The answers to many puzzles in logic and to practical problems can also be found by applying algebra.

■ DISCOVERY ■

Girolamo Cardano

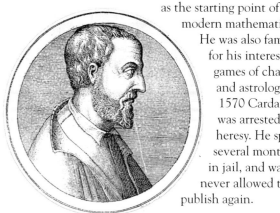

Cardano (1501–76) was an Italian mathematician and physician. In 1545 he published a book, *Ars Magna* (Great Art), in which he described how to solve biquadratic and cubic equations – those where the highest power of any term is 4 and 3 respectively (p. 40). This work is often seen as the starting point of modern mathematics. He was also famed for his interest in games of chance and astrology. In 1570 Cardano was arrested for heresy. He spent several months in jail, and was never allowed to publish again.

⬛ Puzzle

A friend said today that on his birthday, his mother's age was three times his, but in 15 years, she would be twice as old as he. How old are the boy and his mother today? (See simultaneous equations, p. 72. Answer on p. 186.)

_____ EXPERIMENT _____

Showing patterns

By using pencils to make lots of squares, you will see a pattern emerging in the number of pencils needed to form the line of squares. Try to find the formula that defines the number of pencils needed to make a certain number of squares in this pattern. Let us say that the number of squares is represented by the letter S, and the number of pencils by the letter p. How can you link the letters S and p together in a formula? It will help if you think of the formula in terms of $p = ?S + ?$.

YOU WILL NEED
● colored pencils

1 TAKE FOUR OF the pencils and arrange them in a square. Join another square onto the first one, and write down the number of pencils added. It may also be helpful if you note the total number of pencils used up as you go along.

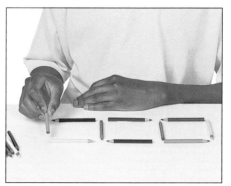

2 CONTINUE MAKING the sequence of squares. Rather than using up all your pencils, can you see a pattern for the number of pencils you need for, say, 10 squares? One way of working this out is to draw up a chart with a column for the number of squares and a column for the pencils. This will help you to write an algebraic formula for the pattern. You can then calculate how many pencils you would need for 300 or 3,000 squares. (Answer on p. 186.)

Calculus

Calculus is a separate area of mathematics from algebra, and one in which algebraic formulas play a vital part. Calculus is a complex study that allows motion and other changes to be measured by "freezing the frame" at one instant in time. To do this, simple concepts such as the rate of change and function are employed. For instance, calculus is used to chart acceleration, which is the rate of change in velocity (speed) over a set time (p. 28). It is also applied to find the maximum or minimum values of an equation or function (p. 73), the area under a curve, or the volume of a particular shape. Calculus is now an important tool in all the sciences, as well as in other technical fields. Engineers, for example, need calculus for determining the performance and strength of structures and engines, and economists use it to analyze rates of change in market performance.

Sir Isaac Newton

In 1687 Newton (1642–1727) published his Principia, often regarded as the greatest written work of science. In it he defined the principles of calculus, as well as presenting the foundations of astronomy and physics in the language of pure geometry. Although Newton had been working on the principles of calculus for some years, Leibniz (p. 47) had published his work on calculus in 1684. At the time there was bitter debate over who had invented it, and the arguments raged for almost a century. Both men must take credit for its invention, but it is Leibniz's notation, with the expressions dx and dy, that is still used today.

Working out the rate of growth

Calculus is used in science to find rates of change. For example, plants grow at an imperceptible rate. But one plant's growth can be plotted over time and analyzed to give a function (p. 73) linking time and growth. It is then possible to give the rate of growth at any time and predict the growth of similar plants in the same conditions. Calculus can also reveal the fastest, biggest, slowest, or smallest instance of a function, such as the maximum height of a ball thrown into the air. At this height the ball will change from upward to downward velocity, so its velocity will be zero. Using this value, calculus can be applied to find the maximum height from any starting point.

DEMONSTRATION

Algebraic chess

By imagining that the board is a grid, you can use coordinates (p. 74) to describe the position of each piece during a game of chess and to record your game. The left edge is the y axis and the base is the x axis. Make a mental note of the axes, or mark them on strips of paper stuck along the edges. You can use this process as a visual aid in solving problems with algebra.

If one piece takes another, the coordinates will be the same

Playing

Set up the chessboard in the usual way. Use just one set of axes, with the white pieces at the bottom of the board. The position for x=1 and y=1 is the bottom left-hand square. Play for a few moves, then write the new coordinates for each piece, noting whether the piece is black or white.

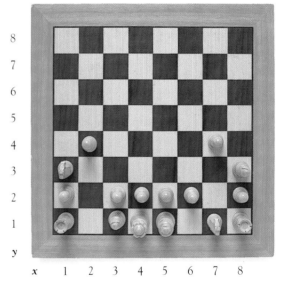

The algebra

On this board, four pieces have been moved. The pawn originally at (2,2) has moved up 2y to (2,4), while the one from (7,2) has moved to (7,4). The knight that started at (2,1) has moved up two places (+2y) and back one (– x) to (1,3). One bishop has moved + 2x+2y, from (6,1) to (8,3).

123 Trick

1. Ask your friend to think of a number. Call it x.
2. Find x + 2.
3. Now find (x + 2) × 3.
4. Subtract 6 from this total.
5. Divide by 3. What is $\frac{3(x+2)-6}{3}$? (Answer on p. 186.)

Finding a solution

AN ALGEBRAIC EQUATION is like a pair of scales pivoted on the = sign. To make both sides balance, whatever is done to the left side must also be done to the right. Well-known examples include Einstein's equation for the relationship between mass and energy, $E = mc^2$, where E is energy, m is mass, and c is the speed of light in a vacuum. Among the most widely used equations are linear, quadratic, and cubic equations. Linear equations, such as $x + 2y = 5$, are the simplest, and use terms to the power of 1 (p. 41). Quadratic and cubic equations use terms to the power of 2 (x^2) or 3 (y^3). To solve equations with more than one variable, using, say, both x and y, you need to group them as simultaneous equations — a set of linear equations with values that will work in both.

■ DISCOVERY ■
Fermat's theorem

Pierre de Fermat (1601–65) was a French lawyer with a passion for mathematics. His "last" theorem states that there are no numbers x, y, z that fit the equation $x^n + y^n = z^n$, in which n is greater than 2. Fermat did not leave a proof, and the problem still fascinates academics. Recently, however, a proof has been published.

INVESTIGATING NUMBERS
Simultaneous equations

If you looked into the ring in a dog show, and found that between the dogs and the human owners there were 35 heads and 94 feet, how would you find out how many dogs and humans there are?

1 Make d the number of dogs, and h the number of humans. Dogs and humans have one head each, and there are 35 heads in all.

$d + h = 35$

So $h = 35 - d$

2 Dogs have four feet and humans have only two, and there are 94 feet in all.

$4d + 2h = 94$

3 Substitute the value for h into the second equation.

$4d + 2(35 - d) = 94$
$4d + 70 - 2d = 94$

4 The equation can be simplified to give the value for d.

$4d - 2d = 94 - 70$
$2d = 94 - 70$
$2d = 24$
$d = 24 \div 2$
$d = 12$

5 Substituting the value for d into the first equation, we find the number of humans.

$h = 35 - 12$
$h = 23$

EXPERIMENT
The Pythagorean theorem

The Pythagorean theorem concerning the length of the sides of a right-angled triangle is a famous example of an algebraic formula. The formula is $a^2 + b^2 = c^2$, where c is the hypotenuse (the side opposite the right angle, which is always the longest side). By making squares with sides equal to each side of a right-angled triangle, you can show how Pythagoras' theorem works.

YOU WILL NEED
● *ruler* ● *pens* ● *scissors*
● *graph paper*

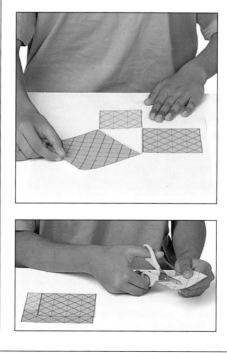

1 DRAW A right-angled triangle on a piece of graph paper. Name the sides a, b, and c (c being the hypotenuse). Measure the length of side a. On another piece of paper, draw a square with sides the same length as a. Repeat this for b and c. Color the squares, then cut them out. Fit them beside the appropriate sides.

2 LAY THE TWO smaller squares of sides a and b on top of the large square. You will have to cut them up to fit them. They should cover it perfectly, so demonstrating Pythagoras' theorem.

DEMONSTRATION
Function machine

A function, often written as f(x), is another way of writing a formula; for example, f(x) = 3x − 7 is the same as writing y = 3x − 7. In this experiment, a simple function machine operated by you and a friend can provide an understanding of how functions work. You mail a number through to your friend, who applies a hidden function and mails the answer back. You must try to guess the function.

1 WRITE a number on a square of poster board. Here, the first number is 3. Draw an arrow going from the number to the space where the answer will be. Mail the card through the top of the function machine.

2 YOUR FRIEND in the machine will apply the function mentally, write down the answer, and send it back to you. The square now has two numbers on it — your original, and the result after the function was applied. In this demonstration, the numbers are 3 → 7.

3 AFTER YOU HAVE received back a number of answers for which the same function has been applied, you may be able to see what the function is. Our numbers are 3 → 7, 4 → 9, and 5 → 11. Can you see the function that has been applied each time? (Answer on p. 186.)

INVESTIGATING NUMBERS
Finding a birthday

We all remember birthdays, but not always the birth day! There is a way to find out the weekday of your birth without using a calendar or asking your parents. If you learn the algebraic method given below, you can try the trick on your friends.

1 First, let y equal your year of birth. We have used the date 22 March 1984 as an example.

$y = 1984$

2 Let d equal the day of the year when you were born. To find d, add the number of days in all the months so far. Note that February has 29 days in leap years (p. 45).

1984 is a leap year

$d = 31 + 29 + 22$
(Jan + Feb + Mar)

$d = 82$

3 Find f using the formula given. Ignore any remainder and work only with the whole number.

$f = \dfrac{y-1}{4}$

$f = \dfrac{1984-1}{4}$

$= \dfrac{1983}{4}$

$= 495 \text{ r } 3$

4 Apply the formula for b. The remainder in the answer will give the number for your birthday.

$b = \dfrac{y+d+f}{7}$

$= \dfrac{1984+82+495}{7}$

$= \dfrac{2561}{7} = 365 \text{ r } 6$

5 Look at the remainder and use this table to find the day of the week of your birth.

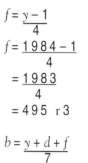

Fri	Sat	Sun	Mon	Tues	Wed	Thur
0	1	2	3	4	5	6

🧩 Puzzle

How many years old is a person if her age now is equal to her age in 3 years' time × 3, less 3 times her age 3 years ago? (Answer on p. 186.)

Cartesian coordinates

THE USE OF ALGEBRA in geometry is now part of science, technology, and data analysis. The concept of geometric algebra dates back to about 300 B.C., when Euclid (p. 114) used a form of it for proofs in his *Elements*. However, the greatest steps were taken by Descartes and Fermat (p. 72) in the second half of the 17th century. With their methods, a point could be represented by a pair of numbers giving its distances from two axes. Cartesian coordinates, named after Descartes, use axes at right angles with an origin (shown as O) where the axes cross. When writing coordinates such as (2,3), the first number represents the distance along the horizontal (x) axis, and the second shows the distance along the vertical (y) axis. Equations can be used with coordinates to plot shapes: if an equation has two variables it makes a two-dimensional shape; if it has three, the shape is three-dimensional. Cartesian coordinates can be used to analyze curves; for example, in a speed-time graph the area under the curve shows distance traveled in a certain time, and the gradient at any point shows acceleration. They can also help to solve simultaneous equations. The point(s) at which the equations' lines cross gives their numerical solutions.

■ DISCOVERY ■
René Descartes

Although he is most famous as a great philosopher, the Frenchman Descartes (1596–1650) was also a scientist and mathematician. In his book *Discourse on Method* (1637), he applied algebra to geometry, and vice versa. The term "Cartesian coordinates" is taken from his name. But Descartes never used a rectangular grid to locate points, so never knew the exact system of Cartesian geometry that we use today.

EXPERIMENT
Coffee temperatures

If you like milk in your tea or coffee, and you like your drink to be as hot as possible, do you put the milk in as soon as the boiling water has been added, or do you wait until you are ready to drink it, then add milk? Surprisingly, Cartesian coordinates can answer that question. Draw a graph. Write "time elapsed after adding water" under the x axis, and mark the axis in 30-second intervals up to 10 minutes. Draw a vertical line on the graph at the five-minute mark. Write "temperature of the liquid" beside the y axis, and mark the axis in 10° intervals, making sure that the scale reaches 212° F (100° C).

Rest the end of the thermometer in the middle of the liquid when taking readings

YOU WILL NEED
● *ruler* ● *pens*
● *sugar*
thermometer
● *stopwatch*
● *graph paper*
● *cup of black
coffee* ● *milk*

1 PREPARE your graph. Set your stop watch to sound every 30 seconds. Next, pour the hot coffee into the cup. Take the temperature of the coffee immediately after it is poured, and plot the point on the graph. Repeat this step every 30 seconds for the next five minutes.

INVESTIGATING NUMBERS
Drawing equations

Equations can be shown as shapes if they are plotted on a graph with x and y axes. The graph also has an origin (O), where the axes cross, at (0,0). Fermat (p. 72) was the first to show equations in this way. He discovered equations for the shapes shown on the right, and for the hyperbola.

1 To plot the equation $x + y = 0$, let x equal a number (here 2). Use this number to find y and solve the equation.

$x + y = 0$
If x is 2
$y = 0 - 2 = -2$

2 Put an × on the graph at the point where $x = 2$ and $y = -2$. Minus values are left of the origin on the x axis, and below the origin on the y axis.

Plot $x = 2$ and $y = -2$

3 Repeat this step, making x equal another number (such as –2), and find y. Plot that point on the graph.

If $x = -2$
$y = 0 - (-2) = 2$
Plot $x = -2$ and $y = 2$

4 Because the equation $x + y = 0$ is linear (p. 72), the graph will be a straight line. Draw a line connecting the points, to show the geometrical equivalent of $x + y = 0$. The equation will fit all coordinates on this line.

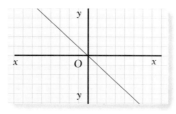

Circle
The equation for any circle is $x^2 + y^2 = r^2$, where r is the radius. In the circle drawn on this graph, $r = 4$; the graph represents $x^2 + y^2 = 16$.

Ellipse
The equation is $x^2/a^2 + y^2/b^2 = 1$, where $2a$ is the length of the major diameter and $2b$ is the length of the minor diameter. Here, $a = 6$ and $b = 3$.

Parabola
The graph shows a parabola that has the equation $x^2 = 4ay$. The y axis is the axis of the parabola, a is the distance from the focus to the x axis, and $4a$ is the length of a chord passing through the focus at 90° to the y axis.

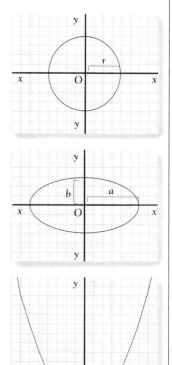

2 POUR IN THE MILK after you have noted the temperature at five minutes. Check the temperature again every 30 seconds for the next five minutes and plot the points on the graph. Link all the points with a smooth line.

Note the temperature

3 REPEAT THE EXPERIMENT with a cup of coffee to which the milk has been added right after the coffee. Take the temperature every 30 seconds for 10 minutes. Use another color to plot the points. According to the graph, which system is better for keeping coffee hot? (Answer on p. 186).

Puzzle
Draw a grid with 40 squares across the bottom and 40 up the side (1600 squares). Plot the coordinates to discover the subject of the drawing. Put a dot at the intersection of the first two points. The picture has a number of separate parts, so do each section separately before moving on to the next.

Part 1: (0,0) (2,8) (4,16) (8,24) (12,32) (16,36) (20,38) (24,38) (28,36) (32,28) (37,16) (37,8) (40,0).
Part 2: (12,12) (16,5) (20,3) (24,5) (29,12).
Part 3: (16,12) (19,13) (20,12) (21,13) (25,12) (16,12) (20,10) (25,12) (20,8) (16,12).
Part 4: (19,15) (20,14) (21,14) (22,15).
Part 5: (13,22) (16,23) (19,22).
Part 6: (24,22) (27,23) (30,22).
Part 7: (14,20) (16,22) (19,20) (16,21) (14,20) (16,19) (15,20) (16,21) (17,20) (16,19) (19,20).
Part 8: (24,20) (26,22) (29,20) (26,21) (24,20) (26,19) (25,20) (26,21) (27,20) (26,19) (29,20).

STATISTICS

Chance and fortune
Most aspects of our lives are subject to events that may be beyond our control. However, mathematics can be used to make predictions on chance events. For example, both meteorologists and insurance companies may want to know the probability of lightning striking particular areas (left). Economists rely on statistics to help them predict changes in financial markets, which may cause huge amounts of money to be gained or lost (above).

STATISTICS ARE PIECES of data that are gathered and analyzed to provide information. The use of statistics can be both an art and a science. Statistical analysis is important in many areas of modern life; it enables researchers to glean in-depth information about present situations as well as to make forecasts about the future. Statistics have been studied only since the 15th century, but their use has significantly affected the way in which modern societies have developed. The spread of disease, the control of scientific experiments, money and insurance industries, and the stock market are all driven and monitored by highly sophisticated statistical formulas.

MESSAGES IN NUMBERS

WHEN RESEARCHERS NEED TO SUM UP the characteristics or actions of a group, they often apply statistics. When estimating the likelihood of an event happening, they use probability. Statistics and probability are useful in many tasks, from forecasting stock prices and writing insurance policies to designing safety systems and studying sub-atomic particles.

In everyday life, we come across many statistics and probabilities. For example, during elections we often hear how people are likely to vote. Similarly, researchers can tell us how much the average adult earns and how long, on average, we are likely to live. The statistics and probabilities we come across are usually put together by statisticians.

Disaster at sea
Safety planners need to consider the condition of ships, predicted weather, and hazards on the shoreline when they estimate the probability of accidents like this oil spill.

Sampling
When statisticians compile information about a large group of people or things, they do not usually study every person or thing in that group. Instead, they focus on a sample whose characteristics will represent those of the whole group. In general, a large sample represents a group's characteristics better than a small one.

Even when statisticians are not studying people, they call the group that they are researching a "population." The information they collect about the population is known as "data."

Keeping count
Statistics grew from the gathering of data. In this Roman relief, a ship's cargo is being counted.

Averages
The average is one of the most important statistics that we use. In statistics, there are three different types of average, each with a precise meaning: the mean, the median, and the mode. Suppose that the shoe sizes of a sample of nine people were, in ascending order, 6, 7, 8, 8, 9, 10, 10, 10, and 11. The mean shoe size is calculated by adding up all the sizes and dividing them by 9, the number of people in the sample. In this example, the mean is 8.7. The median is the shoe size in the middle of the list in ascending order: 9. The mode is the shoe size that occurs most often: 10.

Spread of data
The mean, median, and mode of a set of data usually — but not always — differ. Statisticians often look at these differences to see how characteristics are spread across a population. In the above example, the mode is bigger than the mean because there is a larger spread of small shoe sizes than large shoe sizes.

Many types of data relating to natural phenomena, such as the height of people or the yield of crops, are spread evenly around the mean, with roughly equal numbers of items on each side. Statisticians say data like this is "normally distributed." The shoe sizes of the sample above were not normally distributed but were "skewed" — spread more widely on one side of the mean than on the other.

The German mathematician Karl Gauss (p. 164) studied the spread of different types of data. He invented the term "standard deviation" to describe this spread. Scientists today use standard deviation to estimate the accuracy of measurements.

Presenting data
Statisticians usually present data in graphs or charts to allow key facts to stand out readily. The two most common presentations are the pie chart and the bar chart. The thickest slice on a pie chart, or the tallest bar of a bar chart, represents the value of data that occurs most often (the mode). Data that is normally distributed (evenly spread) will produce a bar chart that has a bell shape.

Statistics of life
People have gathered and used statistics for thousands of years. The earliest statistics, such as the censuses of the ancient Babylonians, Egyptians, and

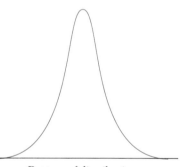

Pattern of distribution
Many groups of data show this pattern. It features a few items at each end of the group, and an upward curve toward the middle, where most items of data occur.

Chinese, were used to count the population for tax purposes. However, from the 15th century onward, statisticians began to realize that statistics could have far more applications.

In the 17th century, an English textile maker, John Graunt (1620–74), decided to study the social problems of his time using statistics. As a hobby, he collected data on deaths in various English cities. He analyzed the data by means of carefully thought out statistical methods. Graunt discovered that what at

A Court for King Cholera
This cartoon was published in Punch, *the British satirical magazine, in 1852. It personifies cholera as an absolute ruler with London in his power. By using statistics to analyze the spread of cholera, people could work out how to prevent it.*

first appeared to be random incidents of suicides, death by disease, and accidents happened year by year in roughly the same numbers. He also found that, overall, more female babies were born than male.

Graunt's work was an early example of epidemiology — the statistical study of health and disease in populations. This science was first used to find the cause of a cholera outbreak in London, England, in 1854. John Snow (1813–58), a doctor, suspected from the way the disease spread that poor sewage and water systems were to blame, so he set out to prove this statistically. He traced the cause of the outbreak to a public well in Golden Square, at the heart of the city. Statistics enabled him to find out how to prevent the spread of cholera 30 years before the germ that causes the disease was identified.

The use of statistics in genetics, the study of inheritance, was developed in the 19th century. This science began when an Austrian abbot, Gregor Mendel (1822–84), used statistics to find out how the characteristics of peas were passed on from one generation to another.

Epidemiology was applied in 1964 to confirm the link between smoking and lung cancer, and is now used to trace the factors that cause other major problems, such as AIDS. Similarly, genetics helps us understand how susceptibility to heart disease, asthma, cancer, and other illnesses is inherited.

Chance
Although almost nothing about our future is certain, statisticians can estimate the likelihood of particular events occurring by means of the theory of probability.

Many of the mathematicians who first developed the theory of probability were gamblers. They hoped that an understanding of probability would improve their chances of winning games. One such gambler, Girolamo Cardano (p. 70), was also a professor of

mathematics. Cardano calculated the probability of throwing certain dice and of pulling aces from decks of cards. He presented his work in his book *Liber de Ludo Aleae* (Book of Games of Chance). In this book, he not only discussed the likelihood of winning fair games but also suggested interesting ways to cheat. For example, he explained how to increase the chances of pulling out a certain card from a deck by rubbing it with soap.

The French mathematician Blaise Pascal (1623–62) was also interested in probability. He worked with his fellow French mathematician, Pierre de Fermat (p. 72), to develop a complete theory of probability. Although their theory concerned frequently occurring events, Pascal was also interested in the probability of freak events occurring. In particular, he wished to find out the probability of miracles happening and devised a special theory of probability for these types of event. We now use the probability that Pascal developed for studying miracles to examine freak occurrences of very different kinds, such as accidents, machine failures, and bouts of severe weather.

Blaise Pascal
Together with Pierre de Fermat, and with the aid of the pattern now called Pascal's triangle (p. 48), Pascal developed mathematical theories of probability.

New ideas
In the last 80 years, a new form of physics, quantum mechanics, has been developed. It is based on the idea that no event, particularly what happens inside an atom's nucleus, can be guaranteed; physicists can only predict the probability of certain events happening. This science demonstrates that statistics and probability are at the heart of our understanding of the Universe.

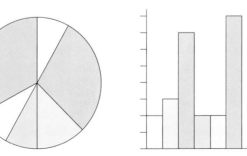

Information in pictures
Pie charts (left) and bar charts (right) are used to organize information so it can be understood instantly. Pie charts show the proportions of different items in one category — for instance, numbers of men, women, and children using a leisure center. Bar charts show different classes of item, such as numbers of people using leisure centers, cinemas, and theaters.

What are the chances?

PROBABILITY IS THE LIKELIHOOD that an event will happen. It is expressed as a number between 0 and 1. 0 signifies that an event can never happen, whereas 1 means that it is certain to happen. A probability of "one in five" can be written as 1/5, 0.2, or 20%. Many researchers in science and industry use probability figures based on past results to predict the future and plan ahead. Some events happen "at random," or by chance in an unpredictable way. The possible outcomes of all such events are equally likely. In other cases, probability cannot be worked out by theory, so a test is done to provide a figure. For instance, a manufacturer of yo-yos may need to check some of them to see what proportion of them are faulty. The bigger the sample, the more accurate the probability will be.

🧩 Puzzle

A bag contains one ball known to be either black or white. A black ball is put in, the bag is shaken, and a ball is drawn out, which is black. What is the chance of the remaining ball being black? The answer is 2/3. One outcome will be drawing a white ball. So why are there two chances of drawing black? (See below.) If a white ball is drawn first, what is the probability of the remaining ball being black (p. 186)?

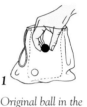

1 *Original ball in the bag is white, and is left inside while the black is drawn out.*

2 *Original ball in the bag is black, and the added black is drawn out again.*

3 *Added black is left in the bag, and the black original is drawn out.*

EXPERIMENT

Guess the shape on the card

You can study probability by using cards marked with simple shapes. With five different cards to choose from, the probability of correctly guessing the shape of an upturned card will be just 1/5, with four it will be 1/4, and so on. Monitor your score on a chart to see if your guess rate is higher or lower than that predicted by the theory of probability. You need to do this at least 25 times to get a realistic result. What do you notice about the accuracy of your guesses?

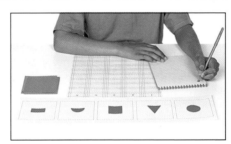

1 **MAKE FIVE SQUARE CARDS** and two sets of shapes. Glue a shape on each card. Lay a duplicate set of shapes on the paper. Make a chart to record your guesses.

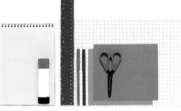

YOU WILL NEED
- *glue* ● *notepad* ● *ruler* ● *pens* ● *scissors*
- *2 sheets of different-colored poster board*
- *large sheet of graph paper*

1	2	3	4	5
✗	✓	✗	✓	✓
✗	✗	✓	✗	✓
✗	✓	✗	✗	✓
✓	✓	✓	✓	✓
✗	✗	✓	✓	✓

The chance result
Add the check marks in each column. The probability of guessing the first shape correctly is 1/5; for the second it is 1/4; for the third, 1/3; for the fourth, 1/2; for the fifth, 1. The expected total number for this experiment of 25 guesses should be 25 × (1/5 + 1/4 + 1/3 + 1/2 + 1) = 57. If the total number of check marks on the chart is greater than 57, this means that you have guessed well.

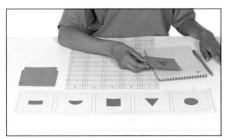

2 **WITH THE CARDS** face down, guess the shape on the first one. Turn the card up and make a check mark for a correct guess or an ✗ for a wrong one.

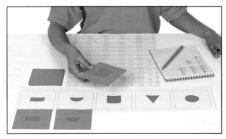

3 **LAY THE CARD** above its symbol. Continue guessing and turning up cards. Shuffle the cards and start all over again. Repeat this at least 25 times.

INVESTIGATING NUMBERS
Lucky birthdays

Some people think that 7 and 12 are lucky. What is the probability that a randomly selected person has a birthday on the 7th or 12th of a month?

1 Work out the probability that a birthday occurs on the 7th or 12th of January. Write down this probability.

$$1/365+1/365$$
$$=2/365$$

2 Work out the probability that a birthday occurs on the 7th or 12th of February. Write this as a fraction and add it to the probability for January.

$$1/365+1/365$$
$$+2/365$$
$$=4/365$$

3 Work out the probability that a birthday will occur on the 7th or the 12th of the other months. Add together the results for all the months to reach the final figure. The probability of someone having a birthday on the 7th or 12th of any month is 24/365 or about 6½%.

$$1/365+1/365$$
$$+1/365+\ldots$$
$$+1/365$$
$$=24/365$$
or 6.575%

The *Titanic*

The British liner *Titanic*, built in 1912, hit an iceberg on its maiden voyage and sank, with the loss of 1,513 lives. The probability of this happening was believed to be extremely remote because of the ship's safety features. Insurance companies work with probabilities, using past claims as a basis for predicting the probabilities of future ones. But because their predictions are only theoretical, they can be wrong.

DEMONSTRATION
Showing probability with a tree diagram

A tree diagram shows the number of possible outcomes or events, with the total probability of each event written at the end of each branch. Here, a boy is randomly choosing an outfit to wear. He has two pairs of jeans, three T-shirts, and two pairs of shoes. First, the probability that he will wear either pair of jeans (1/2) is shown by two branches. For the T-shirts, a further set of branches is added to that for each pair of jeans, and so on. To find the total probability of certain combinations, add up the relevant probabilities along a row, so for outfits without a red T-shirt, 1/6+1/6+1/6+1/6=2/3.

Which jeans?
There are two pairs of jeans, so the total probability that the boy will wear either pair is 1/2. This is shown with two branches.

Which T-shirt?
Each pair of jeans can be combined with one of three T-shirts, so the probability for each is 1/3. The total probability for each combination of a T-shirt and a pair of jeans is 1/2×1/3, which equals 1/6.

Which shoes?
There are two pairs of shoes, so the probability that each set of jeans and T-shirt will be combined with either pair of shoes is 1/2. The total probability for each outcome of jeans, T-shirt, and shoes is 1/2× 1/6, or 1/12. This can be proved by adding the final probabilities on the tree diagram; they will equal 1.

1/2   1/2

1/6   1/6   1/6   1/6   1/6   1/6

1/12 1/12 1/12 1/12 1/12 1/12 1/12 1/12 1/12 1/12 1/12 1/12

Looking at averages

AVERAGES ARE USED DAILY in school reports, business, and industry. People may speak generally of an average family, weight, or height. In statistics the term "average" usually refers to the mean. This figure is equal to the sum of all the measurements taken, divided by the number of items. For instance, the mean weight of oranges in a bowl is their total weight divided by the number of oranges. Sometimes it is not appropriate to use the mean, particularly if the answer does not make sense. For example, in a group of four horses and five riders, the mean number of legs is 2.88. Statisticians also use other types of average — the median and the mode. The median is the value exactly in the middle of a group. The mode is the measurement that occurs most frequently.

Computer software

Some varieties of computer software, particularly those known as spreadsheets, which are a means of manipulating columns and rows of figures, can be used to calculate average increases or decreases across a vast amount of data. Calculations for forecasting that would have taken a person many hundreds of hours to compute on paper can now be performed in just seconds. Computers are used by statisticians to analyze and forecast a whole range of data, from currency exchanges to stocks and shares or the number of hospitals or schools needed in a certain area. In all of those calculations, averages are employed within the process to get a general starting figure on which to base the predictions or analyses.

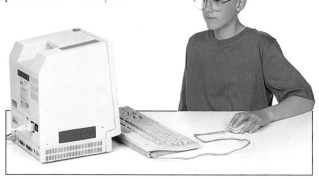

Changes in average height

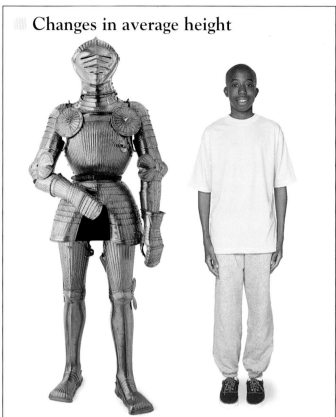

The average height of humans has increased over the centuries. This is due in part to better nutrition, living conditions, and medical care. The picture above shows an average 11-year-old boy in the late 20th century. At 5 ft 4 in (1.63 m), he is almost as tall as the German knight for whom this armor was made in the 16th century.

Curve of distribution

The Prior's Staircase leading to the Chapter House in Wells Cathedral, England, dates from the 13th century. On each step is a worn section — a curve of distribution — that has developed over centuries as people have used the stone staircase. The steps are worn very heavily where people have walked most often. Many collections of data show this pattern of distribution, with the greatest concentration of results always nearest the center.

Mean, median, and mode

The mean, median, and mode of a set of values can be worked out to show different averages. The group of children in the pictures below ranges in height from 4 ft 6 in to 5 ft 4 in. The scale on the left of each picture gives a rough guide to each person's height. In each picture the person or people representing the average are highlighted. Look at the pictures carefully to understand why the highlighted figures represent a particular average for their group. You could find out the heights of a group of your friends and then work out the mean, median, and mode, or consider how these statistics might be used to demonstrate how the average height for 10- to 12-year-old children has changed over the years.

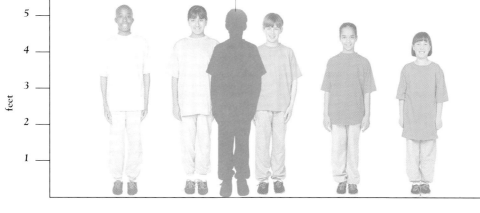

This imaginary figure is the mean height of 5ft

The mean
To find the mean, the heights of all the children are added together and the total is divided by the number of children. Here, the total height (working from left to right) is 5 ft 4 in+5 ft 4 in+5 ft+4 ft 10 in+ 4 ft 6 in, or 25 ft. Dividing by 5 gives 5 ft, shown by the gray figure. If one child was exceptionally tall or short, though, this could distort the mean so that it did not truly reflect the heights of most of the children.

This boy is the median height

The median
This is the measurement at the half-way point of the group. There are just as many items of data above as below this point. It can easily be measured by eye. Here, the median is 5 ft, which is the height of the boy in the blue shirt. Generally, if the mean and the median for a group are similar, it is better to use the mean as the average. However, if they are very different, it may be better to use the median.

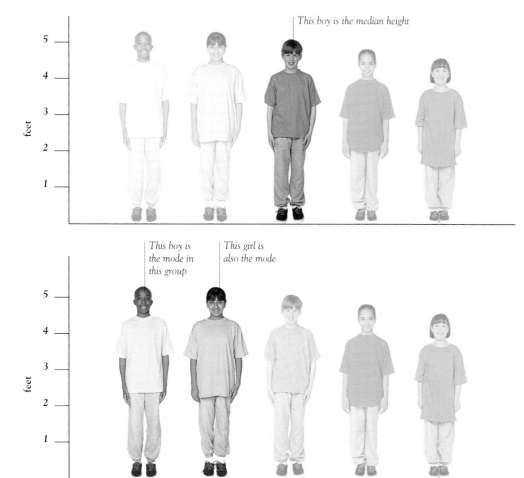

This boy is the mode in this group

This girl is also the mode

The mode
The mode is the value that occurs most frequently in a group. In this picture the boy and girl on the left are the same height, 5 ft 4 in, but all the other children are of different heights. Because this measurement occurs twice, while all the rest occur only once, 5 ft 4 in is the mode. In a group where many of the measurements are the same or very similar, it may be sensible to use the mode as the average.

Presenting data

GRAPHS SHOW NUMERICAL DATA visually. Information given in this way is easier to absorb than words or numbers are. The most common graphs are line graphs, pie charts, and bar charts. These can illustrate the same point in different ways. For example, a line graph may show how bicycle prices change over several years — rising or falling or staying the same. A pie chart may be used for comparing the number of girls' bikes and boys' bikes sold, and a bar chart can show the relationship between bicycle prices in different stores. Graphs are often used to support an argument, but they can also be manipulated to give confusing information.

EXPERIMENT
Ocean floor profile

This "ocean" profile is drawn on graph paper. When you link the points on the graph, you will have an outline of the ocean floor with information about troughs and peaks. Oceanographers plot deep-ocean ridges in this way. They send sonar beams from a ship to the sea bed, and the beams bounce back to a computer on the ship, giving the depth. The computer translates the measurements to make an electronic map.

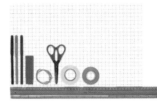

YOU WILL NEED
● *pens* ● *modeling clay* ● *string* ● *scissors* ● *adhesive tape*
● *ruler* ● *graph paper* ● *water-filled fish tank with the bottom covered in stones and other obstacles to resemble the sea floor*

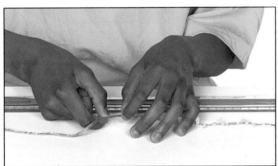

1 CUT A piece of string 12 in (30 cm) long. Wrap each end in adhesive tape. Hold the string against a ruler and mark it at ½-in (1-cm) intervals with thin pieces of tape.

2 THREAD A LUMP of modeling clay onto one end of the string to act as a weight, and tie a knot to hold it in place. Draw the axes of the graph. Make the horizontal axis the distance across the tank, with each square equal to ½ in (1 cm). Make the vertical axis the depth, with each square equal to ½ in (1 cm). Measure the depth of water by resting the ruler upright against the outside of the tank. On the graph, draw a horizontal line to represent the water level, and write 0 beside it. Mark the vertical axis below the water level with negative numbers going down from 0.

Hold the string at the level reached by the water line

3 PLACE THE marked string over the edge of the tank so that the weight rests on the stones on the bottom. Tape a ruler across the length of the tank, aligning the 0 with the edge on your left.

4 STARTING AT THE ½-IN (1-CM) MARK on the ruler, insert the string into the tank until the base touches the stones. Make sure that the string is fairly straight, and the weight rests on the bottom. Count the number of markers immersed in the water, and put an × on the coordinates on your chart. Continue across the tank at ½-in (1-cm) intervals.

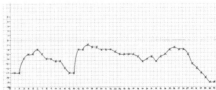

The finished profile
When the floor has been plotted, join the ×s to see a profile of the front of the tank. This picture shows the profile that the boy would see.

EXPERIMENT
A bar chart

A bar chart is a way of presenting data that allows similar items in a group to be compared. The frequency of the different items of data is illustrated by the length of the bars. This experiment shows how to collate information in a three-dimensional bar chart, using blocks of wood. Each of the two dials lands on a number, then the numbers are added together and a block is placed on that total. You will soon see a pattern emerging (p. 79).

YOU WILL NEED
- 16 building blocks
- ruler • scissors
- paper • graph paper • pens
- glue • pencils

1 USE this template to make the dials. Cut the triangles in different colors for each number, stick them onto thick paper (or card), and number them 1, 2, 3, and 4.

EXPERIMENT
What you do in a day

A pie chart is based on a circle. The total area of the circle represents 100 percent of the information, and is divided into segments, each of which represents a category of information. The pie chart is a good way to show how a part of something relates to the whole. It is an effective visual reference often used in newspapers. You can make a pie chart about what you do in a day and see at a glance how your chart might compare to that of a friend or someone in your family.

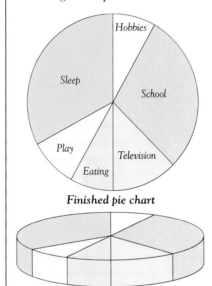
Finished pie chart

Computer-generated chart

Preparing the raw data
Write a list of things that you do every day, such as sleeping, eating, going to school, and watching television. Note the number of hours that you spend on each activity in a typical day. Next, work out the fraction of the day that you spend on each activity. The whole pie (360°) equals 24 hours, so if you spend 8 hours sleeping, that is $\frac{8}{24}$ ($\frac{1}{3}$), or $33\frac{1}{3}\%$, of a day. So the angle of the section will be $\frac{1}{3} \times 360°$, or 120°. Make sure that the figures add up to 360°.

Making the chart
You can make a pie chart by hand, using a protractor, compasses, and ruler, or you can feed the information into a computer, using a spreadsheet, and print it out.

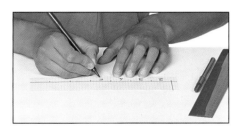

2 MAKE A HOLE through the center of each dial. Insert a small pencil, pushing it through from the top of the square, and move the square along it so that the square is halfway up the pencil. The point of the pencil lead will be the spinning point.

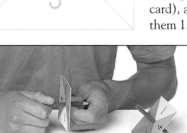

3 ON A PIECE of graph paper, draw a vertical line equal to one block length. Draw a horizontal line, and mark off sections a block-length apart. Write the numbers 2, 3, 4, 5, 6, 7, 8 in these sections. This will be your horizontal axis. Cut off the unused paper.

4 NOW SPIN both the dials at once, one in each hand. Hold each pencil upright and twirl it to start it spinning. When the dials stop and fall over, add together the two numbers on the sides that touch the tabletop. Put a block on the chart by that number.

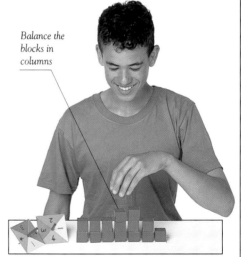
Balance the blocks in columns

5 CONTINUE IN THIS way until all 16 blocks have been placed on the bar chart. What do you notice about the distribution of the blocks? Do the experiment again and see if the results are very similar to those on the first chart or noticeably different.

Information on graphs

DATA GATHERED IN EXPERIMENTS, and plotted on a graph, can be interpreted to show relationships between quantities. For example, does a person's weight increase as his or her height increases? Does an electric current change if the voltage in a circuit is increased? Such relationships can be shown in various ways. On some graphs a line or curve joins all the items of data, to illustrate features such as a change over a set time. On scatter graphs a line is drawn through a group of data to show general tendencies. Scatter graphs are often used to interpret scientific data. They are also necessary in marketing, for the analysis and prediction of sales after a certain project, and in finance and business for predicting price rises in stocks and shares, or supply and demand for particular products.

Recovery rate

These days, athletes use a lot of statistical information when training, to make sure they are in top form in speed, strength, and stamina. Fitness is important for everyone, and one of the accepted ways to discover how fit you are is to see how long it takes your pulse rate to return to normal after vigorous exercise; the less time it takes, the fitter you are. This recovery period can be plotted on a graph. Test your parents, friends, or brothers and sisters as well, to see who is the fittest in your sample group. This experiment tests your level of aerobic fitness (how efficient your heart is).

YOU WILL NEED
● *jump rope* ● *stopwatch* ● *ruler*
● *pens* ● *notepad* ● *graph paper*

1 TAKE YOUR pulse either on your wrist or just under your jaw. To feel for the pulse, use your second and third fingers, not your thumb. Count the pulses over 15 seconds and multiply by four to get the rate for one minute. Write down your pulse rate. This figure is the resting pulse rate.

Use your index finger to feel for a pulse

2 NOW SKIP or do some other brisk exercise for about a minute, and take your pulse again. Continue to check and record your pulse rate every minute until it returns to the resting pulse rate. Draw a graph, showing the time in minutes on the horizontal axis and the pulse rate on the vertical axis.

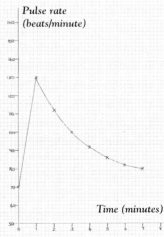

Recovery period
On the chart, every four squares on the horizontal axis represents one minute, and every five squares on the vertical axis represents 10 heart-beats. Note that in this example, the vertical axis starts at 50 rather than zero.

Pulse rate (beats/minute)

Time (minutes)

EXPERIMENT
Line of best fit

In most statistical work and scientific experiments, results can be plotted on a scatter graph to show any relationship between two features of the data. Often, the data will not show an exact relationship, because all results from experiments contain some degree of error, but it will probably indicate a general trend. A "line of best fit" is drawn through the group of crosses plotted on the graph to show the ideal relationship. This experiment shows how to plot the relationship between height and shoe size in a group of children. The information on the scatter graph can then be used to predict the likely shoe size of someone of a certain height.

YOU WILL NEED
● *ruler*
● *notepad* ● *pens*
● *tape measure*
● *graph paper*

The end of the tape measure should align with the top of the friend's head

1 FIRST, COLLECT DATA from a group of your friends. Try to gather information from as many people as possible — tall people and short, people with large feet, males and females, adults and children — so that you have a wide range of statistics. Measure the height of each friend, then measure the shoe length from the heel to the big toe. Make a note of all the measurements.

3 POSITION THE RULER diagonally on the scatter graph so that it passes as accurately as possible through the center of the group of ×'s. Draw a line through the group, making sure there is an approximately equal number of ×'s on each side. This central line is your line of best fit.

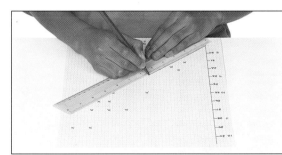

■ DISCOVERY ■
Ludwig Boltzmann

Boltzmann (1844–1906) was born in Vienna. At the start of his career, in the 1860's, the theory of thermodynamics was established, predicting the behavior of gases in certain conditions. The kinetic theory of gases (related to their movement) was also being developed. Boltzmann discovered the statistical links between the two, and founded the theory of statistical mechanics to discover the behavior of large groups of atoms. He committed suicide at 62, depressed at his colleagues' failure to appreciate his work. However, his predictions concerning atomic behavior, based on statistics, were upheld by experiments in atomic physics carried out about 50 years after his death.

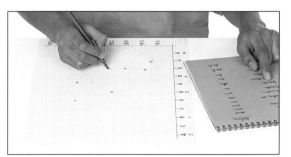

2 DRAW A GRAPH, with shoe length on one axis and height on the other axis. You need not start with 0. Shoe length could start at 8 in (20 cm) and height at 4 ft 3 in (130 cm). Using your data, mark an × on the graph for each person.

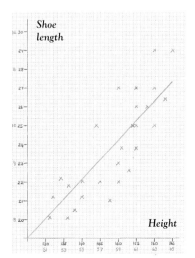

Shoe length

Height

Graph with line of best fit
The ×'s on the graph represent the heights and shoe lengths of your friends. As you can see, the ×'s do not fall in a perfectly straight line, so you have to use your judgment when finding the line of best fit. Sometimes the relationship between features is shown most clearly by a curved rather than a straight line. On some graphs the line appears to change direction or gradient at one point. This is of great interest to scientists, because it indicates that a relationship has changed. For instance, when metal is stretched, it will eventually reach a certain point where it cannot be stretched any farther, and will break.

MEASUREMENT

Precise measurement

Times achieved in sports are measured in hundredths of a second. The difference between the gold and bronze medals in a swimming race (left), for example, might be as little as 0.02 seconds — less than the blink of an eye. Calipers measure the widths of three-dimensional shapes such as stones and screws. This replica from a set of Chinese sliding calipers (above) has a scale on the fixed arm to show the measurement.

WITHOUT SYSTEMS and methods of measuring things, science and engineering would not have progressed as far as they have today. A system of measurement using cubit lengths enabled the building of superb structures such as the Great Pyramid of Khufu at Gizeh, in Egypt, thousands of years ago. Today, atomic clocks are accurate to within billionths of a second. People use measurements every day, from calculations of weight or time, to money transactions dealing electronically in millions of standard units.

SYSTEMS OF MEASUREMENT

Measurements are used to gauge physical features such as length and weight, and to chart the passing of time. Scientists also measure more complex quantities such as the speed of certain objects, the forces acting on them, and the amount of electric current flowing between them. They use measurements to investigate these quantities mathematically.

Almost any quantity in the physical world, such as weight, height, or electric charge, can be measured. When we measure something, we define its properties in relation to special units of measurement. The height of a tree, for example, can be measured in feet or in meters. Over many centuries, people have developed systems of measurement that are accepted in most of the world.

Ancient records
This Egyptian wall painting from the 18th dynasty (1567–1320 B.C.), in the tomb of Menna at Thebes, shows workers measuring a field of grain and, below, recording the crop yield.

▥ Early units
The first units of length were based on body parts (p. 92). Units of weight, volume, and power evolved from commonly used containers or from amounts that a human or an animal could move. Horsepower, for example, a unit of engine power, is based on the power that a horse can exert when pulling an object.

In 221 B.C., the Chinese emperor Shih Huang Ti set standards for the Chinese system of weight measures. A standard vessel for measuring wine or grain was defined not only by its weight but also by the note it made when it was struck. Given a uniform shape and a fixed weight, only a vessel of a certain volume would make a particular note. For this reason, the ancient Chinese words for "wine bowl," "grain measure," and "bell" are the same.

Units of weight developed as people traded with each other. Standard weights, made from metal or stone, were first made by the Babylonians and Sumerians, but soon spread throughout the Middle East. The *mina* was a standard weight commonly used in this area. In some records the *mina* weighs about 22 ounces and in others, 34 ounces. It was divided into 60 smaller units, each of which was called a *siqlu*. A group of 60 *minae* made a larger unit known as a *biltu*.

Coins for trading evolved from fixed metal weights and were often named after the weights from which they originated. The *siqlu*, for example, is referred to in the Bible as the "shekel," and the *biltu* gave rise to the Greek coin called the "talent."

▥ Imperial system
Over the centuries, many systems of measurement have been developed and refined as people have tried to find consistent measures for use in commerce. European traders in the Middle

Standard weights
The metal cylinder on the left is a standard kilogram, one of the modern SI units of mass. Throughout history, a wide variety of different objects, many specially made, has been employed for measuring quantities such as weight. The elephant from Burma and the Ashanti warrior from Africa are examples of traditional standard weights.

Information from stars
An octant measures the altitudes of stars. It was used to find the bearing of a ship or the time at night.

Ages, for example, created a system called "avoirdupois." This was based on the old French words *aveir de peis*, meaning "weight of goods." This measurement system was one of many that gradually became part of the imperial system. The imperial system is based on units such as inches and pounds, which have been in widespread use for more than 750 years. The system was officially introduced in the Magna Carta, the English charter signed by King John in 1215, and was last refined in 1968. Large and small units in the imperial system have very unusual connections, which do not follow any particular pattern. A foot for example, is 12 inches, but a yard is 3 feet, and a mile is 1,760 yards. Moreover, to this day definitions of imperial units vary slightly from country to country. These problems have made the imperial system difficult to work with and have led scientists, engineers, and manufacturers in many countries to adopt the metric system.

▥ A system for all
Today, the metric system is the most common standard measuring system and is gaining popularity. In metric, length is given in meters, mass in kilograms, time in seconds, and current in amperes. Wherever you go in the metric world, you will obtain the same result if you measure an object using metric units. The system is consistent because it is based on special

Volume and power in vehicles
The strength of motor vehicles used to be measured in horsepower. Modern vehicles are described by giving their engine size in liters, a metric unit of volume. This Jaguar E-type, at over 4 liters, is very powerful.

quantities that never vary, such as the properties of certain atoms (p. 108) and of laser light.

The metric system was first proposed in France in 1670. It was designed to fulfil two important requirements: that every unit in the system could be derived from a small set of standard units, and that large units could be made by multiplying small units by 10, 100, or 1,000. The system was not put into use in France until 1799. In 1875, it was also officially adopted by many other countries. The system was refined in 1960 when the *Système International d'Unités* (International System of Units) was introduced. *Système International* (SI) units are now defined using very precise, scientific methods.

At the heart of the metric system is the meter. This unit of length was originally defined as 1 ten-millionth of the distance between Paris and the North Pole. In the 20th century, the meter, along with all of the other measurements in the metric system, has been refined so that it is accurate

Candle clock
This is a variation on an old type of clock. It takes 10 minutes to melt from one pointer to the next.

enough for high-precision scientific measurement. Today, a meter is defined in terms of waves of a certain type of laser light. Scientists today often work at astonishing degrees of accuracy, using units such as the nanometer (which is less than one-billionth of a meter). When they study very large or small quantities, they use specially defined terms called "multipliers." "Mega," for example, is a multiplier of one million, and "micro" is one millionth. So 3 megajoules equal 3 million joules, and 7 microseconds equal 7 millionths of a second.

Areas are calculated by multiplying one length by another (p. 98), so areas in the metric system are given in square meters. Similarly, volume (p. 102), found by multiplying three lengths together, is often measured in cubic meters.

Although most people consider the expressions "mass" and "weight" to mean the same thing, in science the words have distinct meanings (p. 106). For this reason, the metric system defines mass and weight with different units. Mass is measured using the kilogram (the SI unit), while weight is given in newtons. The mass of a particular object is the same wherever it is in the Universe, but its weight varies according to the influence of local gravity.

Measuring by sound
A sonic tape measure aims sound waves at an object at the speed of sound, and calculates that object's distance from the time taken for the waves to return.

Time

Early humans measured time just by day, night, and the seasons. Later people devised calendars that marked time in years. A year was divided into months, based on the waxing and waning of the Moon. The Western, Muslim, and Jewish calendars are still based on the movements of the Sun or Moon. The Western calendar has been refined repeatedly since Roman times, and now gains only one day every 3,200 years.

The metric system uses seconds as the basic unit of time. Some of the most accurate modern timepieces are clocks containing quartz crystals (p. 108). Atomic clocks, the descendants of quartz clocks, are so accurate they gain or lose only 1 second every 1.7 million years.

Other units

The metric system contains many other measurement units. These include the joule for measuring energy, the kelvin and the Celsius scales for measuring temperature, and the candela for measuring the intensity of light. Many metric units are based on other, simpler units. Speed, for example, can be expressed in meters per second. Similarly, joules, which are the units of energy, can be written as kilogram square meters per second squared. The metric system has been put together so that these links can be made readily. However, many people are still reluctant to adopt it for everyday use in countries where the imperial system has been used for hundreds of years.

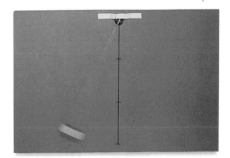

Swinging pendulum
For hundreds of years, the movement in mechanical clocks was regulated by the swing of a pendulum. The time taken for each swing depends on the length of the pendulum.

Personal measuring systems

THE EARLIEST STANDARD UNIT of measurement was the cubit, devised in Egypt about 5,000 years ago and based on the length of a man's hand and forearm. Other systems also used lengths of body parts: the Greeks based theirs on the length of a finger, whereas the Romans subdivided the foot into 12 inches. Feet and inches were also the units of length in the imperial system, introduced in England in 1215. This system spread all over the world, and was widely used until recently. Today, though, it has largely been replaced by the metric system, and by the scientifically defined *Système International* (SI) units (p. 91).

How high is a horse?

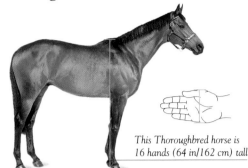

This Thoroughbred horse is 16 hands (64 in/162 cm) tall

Horses are traditionally measured in hands, from the ground to the withers (shoulder blades). A hand is four inches (10 cm), and parts of a hand are given as inches. Ponies are less than 14.2 hands, or 14 hands 2 inches (58 in/147cm), while some heavy draft horses can be more than 18 hands (6 ft/182 cm) tall.

Reach
The reach is the length of a person's outstretched arms from the tip of one middle finger to the tip of the other. On an adult it is roughly equivalent to 2 yds or 72 in (180 cm).

Head
The term "head" is used to describe the difference in height between two people. It is useful for arranging ranks of seating in theaters, where each row of people must be able to see the stage.

Digit
In ancient Egypt a digit was the breadth of a finger. There were four digits in a palm, and 24 in a cubit.

Span
This is the length of an extended hand between the tip of the thumb and the tip of the little finger. It is about 8 in (20 cm). The span was in use in Egypt and Israel at the same time as the cubit, and was equal to half a cubit.

Cubit
This is the measure of the forearm from the elbow to the tip of the middle finger. It is usually between 16 and 20 in (40 and 50 cm).

Hand
The width of a closed hand, from the edge of the palm to the edge of the thumb, has been used for centuries as a standard unit of measurement for horses (see above).

The body

Body parts provided the most convenient units for early measurements of length. The picture here shows the parts of the body from which each measurement was taken. These lengths would actually have been based on the measurements for an adult man, and would have varied widely from person to person. These inexact measures were replaced by systems of international standardized units, but parts of the body are still useful for finding approximate lengths.

Stride
Strides were often used by surveyors and farmers to measure distances along the ground. A stride is about 1 yd (90 cm). With practice, people can make their estimate fairly accurate by adjusting the length of their stride.

Foot
This measure was originally based on the length of a human foot. The standardized measurement is still used today in the imperial system.

Puzzle

How many times would you need to wind a string around your head before it measured the same as your height? (Answer on p. 187.)

The size of Goliath

The Bible (*1 Samuel 17*) gives the exact height of the giant Goliath: "And there went out a champion, out of the camp of the Philistines, named Goliath, of Gath, whose height was 6 cubits and a span." Different types of cubit varied in length, usually according to the personal measurement of the king or emperor of the time. According to the measures (left), Goliath would have been 8 ft 6 in (2.6 m) tall.

EXPERIMENT
Dimensions of a room

Using your own personal measuring unit, you can measure different dimensions in your room. As long as you work consistently with the same unit, you will be able to find the length of the sides and work out the area (pp. 98–99). You can then plan a room layout, for example, when adding or moving furniture.

YOU WILL NEED
- ruler
- tape measure
- pen ● notepad

Measure the length of your shoe from the center of the heel to the big toe

Your personal "foot"
Use the tape measure to find out how long your foot is in standard units such as inches. You can then mark out the dimensions of your room by placing your heel to your toe and counting the number of "feet." Draw a layout on a notepad to use as a floor plan. Then find the real measurements by converting your feet-lengths to inches.

EXPERIMENT
Measuring from fingers to nose

Among textile sellers, an old method for measuring cloth was to hold the fabric between the fingers and stretch it the length of an arm to the nose. This measures about 1 yd (90 cm) on an adult. (It may not be so accurate on a young person.) The method is useful for estimating lengths of string or cloth.

YOU WILL NEED
- a tape measure that has both metric and imperial measurements

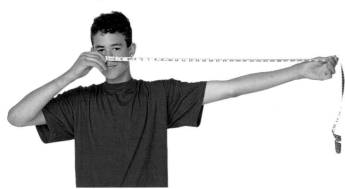

1 TO MEASURE A YARD in the imperial system, hold the end of the tape measure on the tip of your nose, and stretch your arm and hand out as far as you can, just above shoulder height. Pull the tape tight. Look at the point that your fingers reached on the imperial side. The tape should show about 1 yd (90 cm).

2 TO MEASURE A METER, hold the end of the tape measure to your nose and stretch your arm out as before. This time, turn your face away while allowing a little more tape to feed through your extended hand. Look at where your fingers lie on the metric side. This measure should be almost 39 in (1 m).

Measuring instruments

ACCURATE MEASURING INSTRUMENTS are vital for gathering and recording physical data such as volume, area, length, time, speed, light intensity, and many other aspects of the material world. Today, among the SI units (p. 91), a standard weight is still kept for the kilogram, and other measuring instruments are compared against it, but standards for the other units are taken from accurate observations of phenomena such as light wavelengths or vibrations in crystals. Modern instruments range from the ruler to computerized equipment that can show the dimensions of an atom. New devices are being developed all the time to allow further progress. With their help humans can land spacecraft on the Moon, use automatic piloting devices in aircraft, and build tunnels to meet precisely under the sea.

Ancient and modern

Measuring systems throughout history have used measuring devices made from materials that range from stone and brass to plastic. Many of these instruments were based on standard lengths that were kept by people in power.

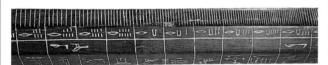

Measuring stick from the reign of Tutankhamun
The Egyptians kept a standard for the cubit against which all official measuring sticks around the country were regularly compared. Its efficiency can be seen from the accuracy of measurements used in the complex construction of the pyramids (p. 154). This one was made from black granite and has numerals and hieroglyphics scratched on to it.

Sonic distance measure
Today, distances can be measured without tape measures or rulers. This sonic distance measure emits an ultrasonic bleep that travels at the speed of sound, bounces off an object, and is reflected to the tape measure. The instrument records the time taken for this to happen, and from this calculates the distance to the object.

A simple ruler

Nowadays, some countries use both the imperial and metric measurement systems in everyday life. A flexible ruler, with both systems of measurement on it, is a simple measuring device that will instantly give conversions from one system to the other, and can be used to measure curved or irregular shapes, such as a plate, a ball, or an egg. This experiment shows how to make a flexible ruler, and how to use it to measure parts of the body such as the wrist.

YOU WILL NEED
● *paper* ● *scissors*
● *ruler* ● *pencil*
● *pens*

1 MEASURE and draw a rectangular strip on the paper, 13 in (32 cm) long by 2 in (4 cm) wide. Make sure that the long sides are parallel. Cut out the strip.

2 USE the ruler to mark inches along one edge of the strip, and centimeters along the opposite edge. Make sure you start with the zeros at the same end.

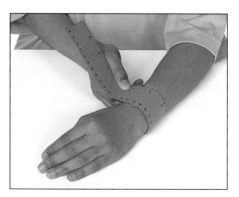

Using the ruler
The ruler is now ready for use. Because it is flexible, it can be used to measure curved as well as straight objects, such as the circumference of your wrist (left). It is also easy to roll up and carry around.

EXPERIMENT
Personal calipers

Calipers are used to measure awkward shapes with curved surfaces, such as spheres, pipes, and screws, which cannot be measured with a flat ruler or tape measure. You can make your own calipers out of poster board, marked with metric and imperial measurements, to keep for use in the future (pp. 96, 104, 155).

YOU WILL NEED
- *ruler* ● *pencil*
- *pens* ● *scissors*
- *adhesive tape*
- *poster board*

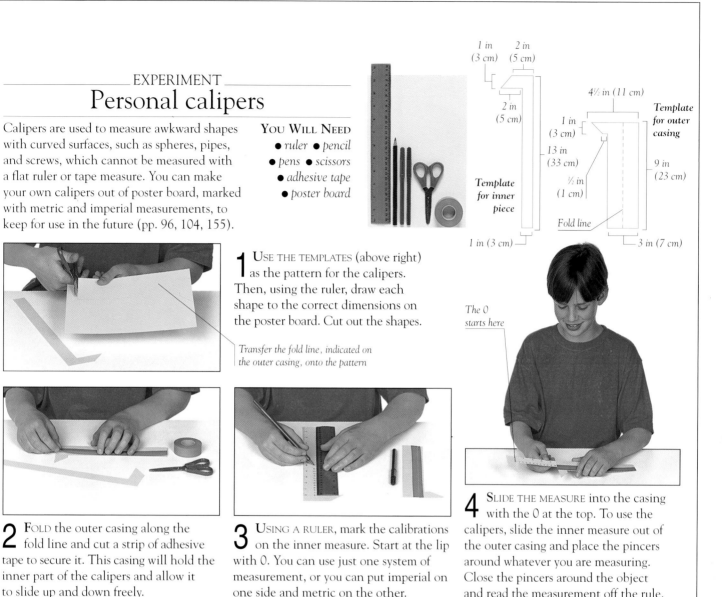

1 in (3 cm) 2 in (5 cm)

2 in (5 cm)

4½ in (11 cm)

Template for outer casing

1 in (3 cm)

13 in (33 cm)

½ in (1 cm)

9 in (23 cm)

Template for inner piece

Fold line

1 in (3 cm)

3 in (7 cm)

The 0 starts here

1 USE THE TEMPLATES (above right) as the pattern for the calipers. Then, using the ruler, draw each shape to the correct dimensions on the poster board. Cut out the shapes.

Transfer the fold line, indicated on the outer casing, onto the pattern

2 FOLD the outer casing along the fold line and cut a strip of adhesive tape to secure it. This casing will hold the inner part of the calipers and allow it to slide up and down freely.

3 USING A RULER, mark the calibrations on the inner measure. Start at the lip with 0. You can use just one system of measurement, or you can put imperial on one side and metric on the other.

4 SLIDE THE MEASURE into the casing with the 0 at the top. To use the calipers, slide the inner measure out of the outer casing and place the pincers around whatever you are measuring. Close the pincers around the object and read the measurement off the rule.

EXPERIMENT
No ruler needed

A unit is a standard for measurement of a physical quantity such as length, volume, mass, or time. Sometimes you can determine these physical characteristics without knowing an object's measurement in standard units. For example, if you want to know the width of a doorway in your house, so that you can move a piece of furniture through it or cut a piece of wood to that length, you can find this out quickly using two overlapping pieces of poster board. By holding them up to the doorway, then against the wood or furniture, you will be able to see how the width of the doorway compares to that of the object.

YOU WILL NEED
● *scissors* ● *pen* ● *poster board*

Measuring and comparing
Cut two pieces of poster board, each more than half the width of the doorway. Partly overlap them, then hold them inside the door frame so that they fill the gap. Mark both pieces with a vertical line where they overlap. Hold the boards against your furniture and realign the line across both pieces, to show the width of the doorway against the object.

Distances

UNITS OF LENGTH, such as feet (p. 90), define distance, depth, and height. Early measuring devices were the human foot and stride. For example, five Roman feet (p. 92) made a pace, and 1,000 paces made a mile. The Vikings measured sea depth in reaches, or fathoms (from *fathmr*, "an embrace"). A weight on a line was dropped to the sea bed, then the wet line was measured in reaches (p. 92). Some modern devices time sound waves (p. 92) or infra-red light beams bouncing off distant objects and convert the time into a distance reading. Other methods involve altimeters (which find height by comparing atmospheric pressure at two locations), radar, and radio waves.

EXPERIMENT
Using calipers

Calipers (p. 95) measure the length and thickness of objects on which it would be hard to obtain accurate measurements with a ruler. They were originally developed to measure the caliber of bullets. The pincers of the calipers are opened, then fitted closely around an object. The distance between the pincers equals that object's thickness or length.

YOU WILL NEED
● *calipers (p. 95)*

The length of your finger is shown between the pincers

Measuring a finger
Hold your finger out straight and line up the top edge of the calipers with your knuckle. Open out the pincers to reach the tip of your fingernail. Read off the measurement.

EXPERIMENT
How far away is the Moon?

You can work out the distance from the Earth to the Moon by using ratios (pp. 54–55). If you make a small counter appear to be the same size as the Moon, the ratio of the actual size of the counter to its distance from your eye will be the same as the ratio of the size of the Moon to your distance from the Moon. If you know the Moon's diameter — it is 2,160 miles (3,475 km) wide — and the size and distance of the counter, then you can calculate the unknown quantity, which is the distance to the Moon.

YOU WILL NEED
● *counter* ● *tape measure*
● *calculator* ● *pen* ● *notepad*

1 STAND outside in a well-lit area, or by a window where the Moon is visible. Hold the end of the tape measure level with your eye. Ask a friend to hold the other end of the tape and position the counter so that it appears to you to fit exactly over the Moon.

2 MEASURE THE DISTANCE between the counter and your eye, then measure the diameter of the counter, and write down these figures as a ratio. Make sure you use the same unit of measurement. The ratio of these two measurements is the same as the ratio of the distance from the Earth to the Moon to the Moon's diameter (see above). (Check your answer on p. 187.)

EXPERIMENT
Finding height with an astrolabe

An astrolabe is an ancient device, dating from about the 6th century A.D., which was used to find the positions of stars and planets. It can also be employed to measure mountains or tall structures, to tell the time, and to navigate. It consists of a flat semicircle marked with a network of lines that show angles of elevation, together with a sighting rod. You can make a simple astrolabe to work out the height of a tree without having to use trigonometry (p. 127). Find the angle of elevation by looking at the top through the astrolabe, then make a scale

drawing on graph paper, using this angle and the distance between you and the tree to create a right-angled triangle.

0° is the vertical line at the center of the semicircle

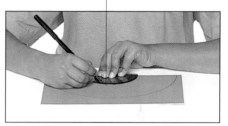

1 ON ONE PIECE OF PAPERBOARD, draw a semicircle with a radius of 5 in (12 cm). Using a protractor, mark angles at 10° intervals — 0° is at the midpoint.

2 RULE LINES on the semicircle to mark the angles. Cut it out. Cut the other piece of paperboard to the same length as the semicircle. Roll it into a tube.

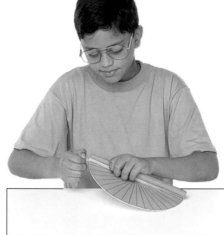

3 SECURE THE TUBE with adhesive tape. Using paper clips, fasten the tube to the straight edge of the semicircle to form the astrolabe's eyepiece.

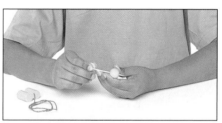

4 TO MAKE THE PLUMB LINE, roll a ball of clay and pierce it with the toothpick. Cut 8 in (20 cm) of string and thread it through. Knot the end.

Drawing to scale
On graph paper, show your relationship to the tree as a triangle. Choose a scale (say 1 square to 20 in/50 cm). Draw a base line to scale, equal to your distance from the tree. From the left-hand end, draw a line at the angle shown on the astrolabe. From the other end of the base line, draw a vertical line at 90° to intersect the angled line. Measure this vertical line and calculate its actual size. Add your height to this figure to find the tree's height.

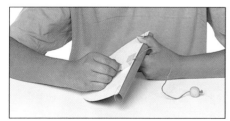

5 SECURE THE PLUMB LINE at the back of the astrolabe, fixing it with tape at the center point. Arrange the string so that it falls over the eyepiece and hangs across the front.

Look at the top of the tree through the eyepiece. Hold the plumb line on the astrolabe, and read the angle

The tree is the vertical line in this right-angled triangle

Note your distance from the tree, in inches or centimeters

The tree's base forms the right angle

Counting squares

AREA IS THE SPACE TAKEN UP by a flat or curved surface. It is measured in units such as square feet (sq ft) or square centimeters (cm^2). In the imperial system, the most commonly used units of area range from square feet to square miles and acres. (An acre is 43,650 sq ft.) The areas of shapes such as rectangles and circles can be calculated with particular formulas. For example, to find the area of a rectangle, multiply the length by the breadth. For a rectangle with sides of 20 in and 30 in, the area is 20×30 or 600 in^2. The areas of more unusual shapes can be found by cutting the shapes to form familiar shapes and using these as a basis for calculations. The boundary around an area is called the perimeter. In some cases, shapes of very different perimeters can enclose the same area. Conversely, shapes that have perimeters of the same length can enclose areas of different size.

Early Italian map

The Romans were prolific builders, especially of public buildings and amenities. This map of Rome dates from the 8th century A.D. It was drawn up from surveyor's measurements. A grid of lines divides the land into squares called *centuriae*, and the territory to be mapped was actually marked out on the ground with special instruments, to create accurate squares. On this map, a grid has been marked on the land between the city walls and the river.

EXPERIMENT

Same perimeter, different area

To estimate an area without doing complex calculations, use a grid and count the squares to find the area. The grid can be a piece of graph paper with small squares (this will be more accurate) or you can rule up your own grid from graph paper with larger squares outlined in red ink as here. The same perimeter can enclose different areas. Using a fixed length of string, guess which shape will enclose the maximum area. (Answer on p. 187.)

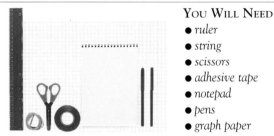

YOU WILL NEED
● *ruler*
● *string*
● *scissors*
● *adhesive tape*
● *notepad*
● *pens*
● *graph paper*

1 TAPE THE PIECE of string so that it forms a loop. Measure it to find the perimeter of any shape it will make.

2 PLACE THE LOOP on the grid in different shapes. Count the number of squares it covers, and note the area each time. (If the shape covers bits of squares, estimate what fraction of each square is covered and add up the fractions at the end.) The perimeter is constant, but how does the area change?

You could make triangles, squares, circles, or irregular shapes such as this one

Puzzle

On graph paper, draw a cross from five squares, as shown below. Cut it out. With two cuts, how can you make the cross into two equal-size squares? Now cut out two more crosses and, with one cut on each, make a bigger square. (Answers on p. 187.)

EXPERIMENT
Same area, different perimeter

In this experiment you can use a handful of crackers to discover how a fixed area can have different perimeters. The crackers have a fixed surface area, but can be arranged to form shapes with outlines of various lengths.

YOU WILL NEED
- pen ● notepad
- square crackers

Making different shapes
Lay the crackers with sides touching, to form a shape of your choosing. Measure the perimeter. You can do this without a ruler, by counting the number of cracker sides that make up the perimeter. Draw the shape and note down its perimeter. Then lay the crackers in a different shape and measure the perimeter again. How does the length of the new shape's perimeter differ from that of the last shape?

EXPERIMENT
Walking through paper

A rectangle of paper can be cut into a shape that you would not believe possible from looking at its original area. In this experiment, you can cut up a small sheet of paper and make a hole big enough to walk through. Show the results of your cutting to a friend, and see if he or she can guess how to do it.

YOU WILL NEED
- scissors
- heavy paper

1 FOLD THE PAPER in half lengthwise, then fold it in half crosswise. Keep folding it crosswise until it is divided into 16 sections. Flatten the creases so that you can see them clearly.

2 STARTING AT one end, cut along the crease from the central fold, stopping ½ in (1 cm) short of the edges. Cut the next crease in the reverse direction, stopping short of the fold. Cut along all the creases, in alternate directions. Then, leaving the end strips intact, cut through the central fold on all the rest.

▨ Spinning silk

The process of making silk from the cocoons of silkworms, illustrated in this 17th-century Chinese painting, is long and arduous. The cocoon is soaked, then unwound and spun into thread. This is twisted and finally woven to make the cloth. The cocoons are only tiny, but each gives a continuous thread of up to 985 yards (900 meters). One square yard of silk uses 10 miles (16 km) of thread.

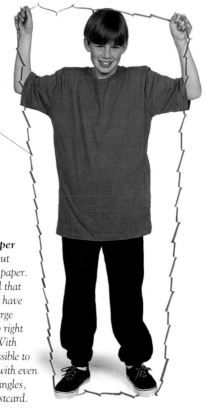

The little sections of paper should all be joined

Stepping through paper
Now open out the string of paper. You will find that the loop you have made is so large you can step right through it. With care, it is possible to do this trick with even smaller rectangles, such as a postcard.

The effect of area

MANY TYPES OF WORK involve calculating areas. Geographers must find the areas of countries or land masses to determine population densities (p. 55) and the extent of mineral deposits. Property developers and real estate agents work out floor areas in buildings to establish business rates and rents. Manufacturers use area calculations to estimate the quantity of raw materials and the cost of packaging they need for their products. Area measurements are also necessary in less direct ways. For example, pressure is calculated as force divided by area. So data on area is crucial in the design of items such as vehicle tires, in which tire pressure can vary with the temperature of the air, the weight of the vehicle, and the area of contact between the tire and the road.

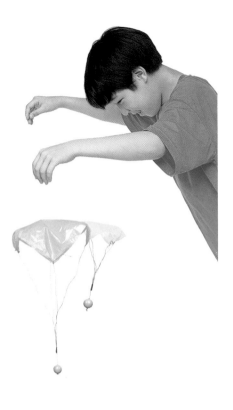

EXPERIMENT
Making parachutes

This experiment shows how the surface area of a parachute affects its resistance to air and its speed of descent. Gravity pulls the parachute down, but as it falls, compressed air pushes against the canopy and slows the parachute's descent. In this experiment you can compare the speeds of descent in parachutes with different canopy areas.

YOU WILL NEED
- *2 plastic bags* ● *ruler*
- *pen* ● *2 equal weights*
- *scissors* ● *string*
- *adhesive tape*

Does area affect speed?
Stand by a window or at the top of a staircase. Hold both parachutes by the center of the plastic and let them drop at the same time. Ask a friend to tell you which hits the ground first. Because the weights and design are assumed to be the same, you will be able to say that any difference in the speed of descent must be due to the surface area of the canopy. All modern parachutes have a hole in the middle to stabilize them. Cut small holes in yours to see if that has any effect on descent speed.

1 DRAW A 12×12 in (30×30 cm) square on one of the plastic bags, and cut it out. Then draw a 6×6 in (15×15 cm) square on the other plastic bag, and cut this square out as well.

2 CUT FOUR PIECES of string 14 in (36 cm) long for the larger square and four pieces 8 in (20 cm) long for the smaller one. Make holes in the corners of each square, and tie a piece of string at each corner.

3 CUT TWO 4-in (10-cm) long pieces of string and tie one end of each to a weight. Using adhesive tape, secure the other end of each string to the four pieces of string on a parachute.

EXPERIMENT
Economy in wrapping

Nearly all commercially sold items need packaging for decoration, to protect the goods during transport, or to preserve perishable products. By wrapping goods using the least possible material, the production costs and waste can be kept to a minimum. This experiment shows how to wrap chocolate economically.

YOU WILL NEED
- *aluminum foil* • *square of chocolate*
- *scissors*

1 POSITION THE SQUARE of chocolate on a piece of the aluminum foil, on the diagonal. Fold in the corners so that they look like an envelope, then trim the foil so that it fits over the chocolate exactly, with no extra foil anywhere.

2 UNWRAP THE CHOCOLATE and reposition it squarely on the foil. Fold in the edges so that they are parallel with the sides of the square. You will easily see which method of folding is the most efficient for wrapping the chocolate.

▮ Computer-aided design

Lasers are used in the textile industry to cut out pieces of fabric for clothing. The computer arranges the pattern pieces for the garment so that the minimum area of cloth is used. This pattern arrangement is then used by the robot when it directs the laser over the material. Computer-aided design helps to minimize the amount of material used, and so reduces costs.

EXPERIMENT
Atmospheric pressure on an area

The Earth's atmosphere is a thin blanket of gases wrapped around the Earth like the skin on an apple, but the pressure it exerts is enormous. Atmospheric pressure can be shown by the equation Pressure = Force/Area. In this experiment, a folded newspaper is laid on a piece of wood. If you hit the wood, the paper will be dislodged. As you unfold the paper and lay it over the wood again, the area will increase, as will the force acting on it. If the force is great enough, it will hold the wood firmly in place, and the wood will break when you hit the end.

YOU WILL NEED
- *cardboard tube* • *lightweight strip of wood or piece of dowel*
- *broadsheet newspaper*
  - *safety glasses*

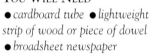

1 PUT ON the safety glasses. Place the wood on the table so that half its length extends over the edge. Lay the folded newspaper over the end of the wood on the tabletop.

2 STAND CLEAR of the wood, then bang the end sharply with the tube. The wood and the newspaper will both fly into the air. Do the experiment again. This time, unfold the paper fully and see what happens.

Occupied space

VOLUME IS THE SPACE TAKEN UP by a three-dimensional object. It is measured in units such as cubic inches (in³), cubic centimeters (cm³), fluid ounces (fl oz), or liters (1 liter = 1000 cm³). Liquids are always measured and sold in volumes, but so are some solids, such as ice cream and coal. Some industries use specific units of volume as their key unit of measurement; for example, wine growers measure by the standard bottle (750 milliliters). In science and engineering, accurate recording of volumes is essential in experiments. Volume is also necessary in working out density: the density of an object is its mass divided by its volume. The least dense solid material known is Seagel, made from seaweed. Seagel is lighter than air and would float away if it were not held down by air trapped in its pores.

▪ DISCOVERY ▪
Handling gases

Joseph Priestley (1733–1804) is best known for identifying oxygen. While working in Leeds, England, he experimented with the gas fizzing off the fermenting vats in the brewery next door. He discovered 10 gases. Priestley's success as a chemist resulted largely from his ability to design effective equipment for his laboratory.

Pig's bladder
Priestley used this pig's bladder as a container for gases, such as the carbon dioxide that he used to make artificial mineral water. It also held gases to be weighed.

EXPERIMENT
The space of your hand

According to Archimedes' principle (p. 62), a body partially or wholly submerged in a fluid experiences an upward force equal to the weight of fluid that it displaces. When Archimedes (p. 18) first formulated this principle, it was fluid displacement that gave him the key. Displacement can also be used to find the volume of a strangely shaped object, which would be hard to calculate arithmetically. In this experiment, you can find the volume of your hand by placing it in a container of water and measuring the water displaced.

YOU WILL NEED
● shallow dish ● measuring jug ● jar ● colored water

1 PLACE the jar in the shallow dish, and fill the jar with water to the top. Gently lower your hand into the water up to your wrist. Water will be displaced and overflow into the dish.

Wait until the water has stopped overflowing before you withdraw your hand

2 POUR the displaced water out of the dish into the measuring jug and note how many fluid ounces or milliliters were displaced by your hand. This measurement gives you the volume of your hand. (One fluid ounce is equal to 1.7 cubic inches.)

Pour the water carefully to obtain the most accurate result

EXPERIMENT

Experimenting with density

The density of a substance is measured in ounces/in³ or pounds/ft³. This experiment shows the way to calculate density, by dividing a substance's mass by its volume. It is not always easy to estimate density. Uranium is extremely dense (1,185 lb/ft³), while balsa wood has a low density (12.5 lb/ft³). In the imperial system, mass is measured in ounces, and volume in cubic inches. Objects whose density is less than that of water will float, and objects whose density is less than that of air will rise. Scientists and engineers often need to know the "relative density" of a material. This is the ratio of that material's density to the density of water (62 lb/ft³). Fresh water has a relative density of 1, while that of sea water is 1.03.

1 To find out the mass of the modeling clay, put it on the scales. Write down the mass in ounces.

2 SUBMERGE THE CLAY in the water-filled jar. Measure how much water is displaced in fluid ounces. To work out the density of the clay, divide the mass in ounces by the volume in fluid ounces.

Does the density stay the same?
The picture on the left shows the modeling clay and the water that it has displaced. Repeat the experiment above, but break off a smaller piece of clay and submerge it in the water. Is the density the same as that of the larger piece? (Answer on p. 187.) What does this tell you about materials and their density? Try with other materials, such as polystyrene or wood.

Measuring engines

Many motor vehicles are driven by internal combustion engines. The engines have cylinders, each containing a piston. Fuel is pushed into the cylinders, then compressed and ignited, and this pushes the pistons up and down to produce mechanical energy. This energy is transmitted to the wheels to make the car run. The volume of displacement of all the pistons determines the power of the vehicle. Some countries, such as the United States, measure engine capacity in horsepower. Every 100 cc equals roughly 10 horsepower (hp).

This powerful Jaguar E-type car has a capacity of 423 hp, or 4235 cc

The floating planet

Saturn, famous for its beautiful rings formed from millions of pieces of ice, is the second largest planet in the Solar System. Although its mass is almost 100 times greater than that of the Earth, it has a relative density of only 0.7 that of water, compared to 5.5 for the Earth. In other words, if it were possible to put the two planets into a Universe-sized bowl of water, Saturn would float but the Earth would sink. The reason Saturn has such a low relative density is that the planet is composed mainly of gases with only a small, solid center.

Working out volume

THE VOLUME OF SOLID SHAPES, or of three-dimensional spaces, is measured in cubic units. There are mathematical formulas for determining the volumes of regular solid shapes such as cones, spheres, cylinders, and cuboids. For these, you need to know measurements such as the length, the width, and the height of each shape. For centuries, mathematicians have been interested in discovering these formulas. One of the first to analyze solid shapes was Archimedes (p. 18). He found that the volume of a sphere was $\frac{4}{3}\pi r^3$, a formula that was inscribed on his tombstone. The volumes of irregular shapes cannot be found this simply. The method used is displacement (p. 102), where the volume is equal to the volume of the liquid displaced by the object.

EXPERIMENT
Formulas for volume

The volumes of different three-dimensional shapes can be calculated with mathematical formulas. Here, the same piece of modeling clay is molded to make three different shapes. As the same volume of material is used, the formulas should all give roughly the same answer. After measuring the shapes, do the calculations to find the volume of each. You will need to know about π (p. 134).

YOU WILL NEED
- *calipers (p. 95)*
- *2 rulers* ● *modeling clay*
- *pen* ● *notepad*

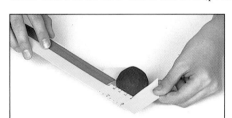

1 MOLD THE modeling clay into a sphere using your hands. Measure the diameter with calipers and divide by two for the radius. Make a note of this figure.

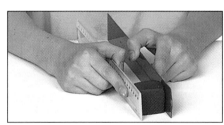

2 CONVERT THE SPHERE into a neat cuboid by squeezing the shape between two rulers. Turn it over, and straighten all the sides until they are flat.

3 MEASURE THE BASE, height, and depth of the new shape with calipers. If they are all the same length, you have made a cube. Note these measurements.

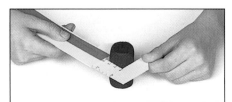

4 ROLL THE CUBE OUT to make a cylinder. Pat it with your hands so that the circular ends are flat. Measure the diameter of the cylinder and divide by two for the radius. Measure the height.

The sphere
The formula for the volume of a sphere is $\frac{4}{3}\pi r^3$.

The cuboid
The formula for the volume of a cuboid is base × height × depth.

The cylinder
The formula for the volume of a cylinder is $\pi r^2 h$, where h is the height.

Cone and cylinder

Does 1 + 1 = 2 ?

Mathematicians know that $1 + 1 = 2$. However, this is not always so in chemistry. This demonstration shows why. Mark the top of a glass. Fill the glass with water to this level, and pour two glassfuls into a larger container. Mark the water level with a strip of tape, then empty the container. Now fill the container with one glass of water and one of rubbing alcohol. The liquid will not quite reach the mark because the molecules of rubbing alcohol are small enough to slip between the water molecules, making a slightly smaller combined volume.

In this experiment you can discover practically how the volume of a cylinder is related to the volume of a cone that has the same radius and height, by filling a cardboard cone with sand. A cylinder with height h and radius r has a volume given by the formula $\pi r^2 h$, while the volume of a cone is expressed as $\frac{1}{3}\pi r^2 h$. Simply by looking at the two formulas, you can see that three full cones of sand should fit into the cylinder exactly.

YOU WILL NEED
● *ruler* ● *saucer* ● *dry sand* ● *pencil* ● *spoon* ● *pair of compasses* ● *adhesive tape* ● *scissors* ● *thin poster board*

× indicates that rubbing alcohol is not for drinking

Are they the same?
If you have done this accurately, the volume of the combined liquids will be very slightly less than that of two glasses of water.

1 WITH THE COMPASSES, draw a circle with a radius of 4 in (10 cm). Rule a straight line through the center to make a diameter. Cut out half of the circle.

2 BEND THE SEMICIRCLE of poster board at the midpoint of the diameter, then fold it around on itself so that it forms the shape of a cone. Secure the cone with a piece of adhesive tape.

Measure the height of the finished cone

3 USE A RULER to measure the height of the cone. Place the end of the ruler flat on the table, with its edge touching the base of the cone. Make a note of this figure.

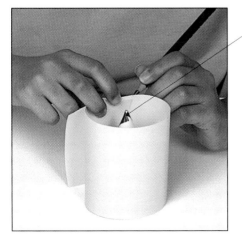

Make a pencil mark where the end of the strip meets the main part

Measure how many cones of sand are needed to fill the cylinder

4 CUT A STRIP OF POSTER BOARD as wide as the height of the cone. Fold the strip around the cone to make the cylinder, with the circumference of the base the same as the circumference of the base of the cone. Mark with a pencil where the strip is to be joined, and secure the edges with tape.

5 PLACE THE CYLINDER on the saucer. Fill the cone with sand and pour the sand into the cylinder. Continue until the cylinder is full of sand. Did you need three cones of sand to fill the cylinder?

Mass and weight

IN EVERYDAY SPEECH the term "pound" is used to describe weight. However, scientifically speaking, the pound is a unit of mass. The mass of an object is the amount of matter in that object. Weight is the downward force of the object, and is defined by the formula: weight = mass × gravity. In science, weight is measured in kgf (kilogramforce) or newtons. An object of a certain fixed mass has different weights on different planets and satellites according to the gravitational field of each body. Astronauts on the Moon, for example, weigh only one-sixth of what they weigh on the Earth. The mass of a group of objects is often used to count the number. For example, if you take a pile of coins to the bank, the teller is more likely to weigh them than to count them out individually. If the mass of one coin is known, then the total number of coins can be determined using special scales.

■ DISCOVERY ■
Albert Einstein

Albert Einstein (1879–1955) was a German-born American physicist who developed the theory of relativity, and with it the idea of the equivalence of mass and energy, expressed in his most famous equation: $E = mc^2$. He showed that the mass of an object increases as its speed increases, but that nothing can travel faster than the speed of light because its mass becomes infinite. Despite a poor educational beginning, Einstein was recognized as one of the most creative intellects in human history.

🧩 Puzzle

Place a large jug on a set of scales and turn the scales back to 0. Now pour water into the jug until you reach the 2¼ lb (1 kg) mark on the scales. Measure the volume of water you have in a measuring cup. What do you find? (Answer on p. 187.)

▥ Paper money

The Chinese were the first to use paper money. From the 10th century, coins in China were made from heavy iron. A thousand coins would have had a mass of about 7½ lb (3.5 kg) — far too much to carry easily. So people began to leave their coins with merchants and use receipts for the coins as currency. These receipts eventually became the official currency. In itself the paper was worthless, but the authorities recognized its value and this gave it worth.

🧩 Puzzle

Do you know which has the greater density (p. 103) — milk or cream? To see for yourself, put a small container on a set of scales and set the scales to 0. Fill the container with milk and note the mass. Weigh the same volume of cream. Did you surprise yourself? (Answer on p. 187.)

EXPERIMENT
How aircraft fly

When an airplane is in the air, the combined weight of its metal, fuel, passengers, seats, and cargo all exert a pull downward. However, the effect of the airplane's total weight is overcome by an upward force referred to as lift, which allows the airplane to fly. Lift is produced by the difference in the air pressure over the upper and lower sides of the wings. The amount of lift generated depends on the airplane's speed of travel and the surface area of the wing — so large aircraft such as jumbo jets, and slow aircraft such as gliders, have very large wingspans. This experiment shows how the lift of an aircraft wing counteracts the weight, allowing the aircraft to fly through the air.

You Will Need
- hair dryer ● scales ● poster board
- ruler ● pencil ● modeling clay
- 2 straws ● adhesive tape ● scissors

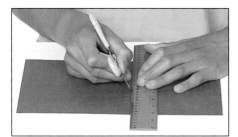

1 ON THE POSTER BOARD, draw a rectangle 8×12 in (20×30 cm). Cut the shape out carefully. Draw a line, 5 in (13 cm) in from one end and parallel to the short edge of the rectangle. Score along this line with a pointed object (the blunt edge of the scissors will do).

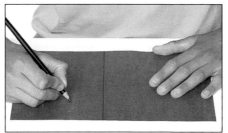

2 WITH A PENCIL, mark two points on either side of the line, each 2 in (5 cm) from the line and 2 in (5 cm) in from the edges of the rectangle. Using a sharp object, such as the point on a pair of compasses, make small holes in the rectangle at these four marks.

3 ENLARGE THE HOLES with a pencil so that a straw can fit through them. As you enlarge each hole, put modeling clay behind the hole to prevent the point from stabbing your finger. Make the holes just big enough to let the straw slide through easily.

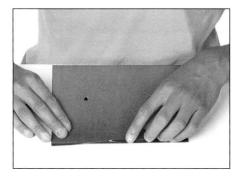

4 FOLD THE POSTER BOARD so that the two short edges are together. Secure these edges with tape. The board will form an arched shape, like that of an aircraft wing. Push the straws through the holes as shown. Wrap tape around them above and below the wing, to hold it in place.

5 ASSEMBLE THE WING on the scales as shown, using modeling clay and adhesive tape. Aim the nozzle of the hair dryer at the front of the wing. Watch the scales to see what happens when your wing "takes off." Try experimenting with wings of different shapes.

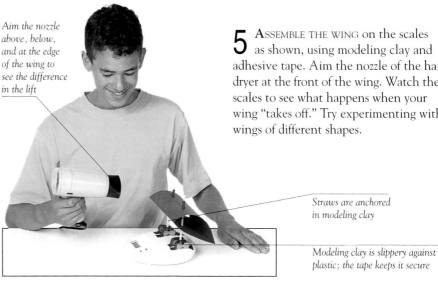

Aim the nozzle above, below, and at the edge of the wing to see the difference in the lift

Straws are anchored in modeling clay

Modeling clay is slippery against plastic; the tape keeps it secure

Air moves faster over top edge

Air under wing has higher pressure than that above

The shape of a wing
A wing is an aerofoil — a shape that enables an aircraft to travel easily through air. Its curved shape forces the air passing over the top edge to travel farther, and faster, than the air passing beneath. This causes the air pressure above the wing to decrease. Because the pressure beneath the wing is higher than that above, it pushes the wing, and therefore the aircraft, upward.

Time

TIME IS MEASURED IN seconds, minutes, and hours. The first timepiece, developed thousands of years ago, was the sundial. The shadow that the Sun cast on a vertical needle showed the time on a disc marked with numbers. Developments in timekeeping have progressed, through water and sand clocks, pendulums, and crystals. Quartz crystals vibrate 100,000 times per second. The timing of these vibrations is so regular, a quartz timepiece is accurate to one second over one month. (A second is now defined as "the duration of 9,192,631,770 time periods between two radioactive states of the atom cesium-133".) These precise time measurements are needed in areas as diverse as nuclear physics and top-level sports.

EXPERIMENT
Candle clock

Many natural elements and forces have been used to tell the time — water, the Earth's movements, and fire. Monks used candles; the wax burned at a constant rate, which helped them regulate their days of prayer and meditation. To make a candle clock, you must first time how fast the wax candles burn, to set the standard.

YOU WILL NEED
- ruler • saucer
- 2 identical candles
- toothpicks
- stopwatch
- matches

Adult help is advised for this experiment

1 TO DETERMINE the rate at which the wax burns, stand one candle in the saucer and light it. Use the stopwatch as you allow the candle to burn down for 10 minutes. Blow out the candle. (Take care not to leave it smoldering.)

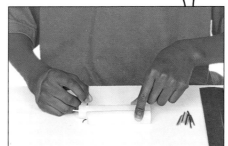

2 MARK THE burnt candle's length on the new one. Using a ruler, measure the difference in length between the two candles. Make a note of this measurement.

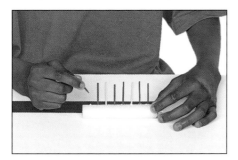

3 ALIGN the tip of the candle (excluding the wick) with 0 on the ruler. Insert toothpicks at measured intervals down to the base, for the 10-minute stages.

This illuminated line is the Greenwich Meridian. It is an imaginary line that runs from the North to the South Pole, through the center of the former Royal Observatory at Greenwich, England. This line was adopted at an international conference in 1844 as the standard to which all other times were compared. In theory, time zones change by 1 hour every 15 degrees of longitude from this point, although countries decide individually what their time zones will be.

Candle clock
Stand the finished candle clock in a saucer. It is now ready to light. As each toothpick drops into the saucer you will have a record of the time elapsed since you lit the candle. Use your watch to check the accuracy of this clock. (Do not leave the candle unattended during this experiment.)

Each toothpick that falls represents 10 minutes

EXPERIMENT
Reaction time

Race drivers and aircraft pilots must be able to react to emergencies in split seconds. You can judge reaction times by making and using this ruler, which shows times from 2 to 21 seconds. It demonstrates that the distance traveled from the time the ruler is released is proportional to the square of the time that it took to catch it.

YOU WILL NEED
- *12-in (30-cm) ruler* • *pen*
- *scissors*
- *adhesive tape* • *paper*

Making the ruler
There are two ways of making the ruler. You can photocopy the template drawn here, enlarging it by 100 percent, and tape it to the ruler. Alternatively, you can cut a strip of paper 12×½ in (30×2 cm) and transfer the second marks onto it from the template, doubling the measurements between the lines to fit your ruler. Then tape the paper to the ruler with the base line aligned on the 0.

Template for reaction ruler

The point where the thumb rests indicates the reaction time

Testing reaction times
Hold the ruler just above a friend's thumb and forefinger. Drop the ruler without warning. The mark where she catches it will show the number of seconds she took to react.

EXPERIMENT
Swinging pendulums

This experiment shows how accurate a pendulum can be for timekeeping, and is derived from Galileo's studies (*c.* 1582) of a swinging lantern. The length of a pendulum is measured from the fixed point to the bob's center of gravity. A pendulum with a period (time for one swing) of 1 second will be 36 in (90 cm) long.

YOU WILL NEED
● *straw* ● *ruler* ● *pens* ● *string* ● *colored adhesive tape* ● *stopwatch* ● *2 spools of thread* ● *modeling clay* ● *scissors* ● *poster board 39 × 32 in (100 × 80 cm)*

1 LAY THE poster board horizontally. Draw a line down the middle and mark it at 10-in (25-cm) intervals. At the top of the line, 1 in (2.5 cm) below the edge of the board, make a hole with the pencil. (Make sure the straw can fit through it.)

2 WRAP COLORED adhesive tape around one of the spools of thread, to make it more visible and add a little extra weight. Then tie a piece of string that is at least 39 in (1 m) long to the spool, threading it through the hole.

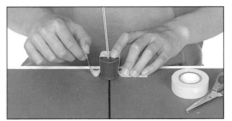

3 ALIGN THE HOLE in the second spool over the hole in the poster board. Push the straw through both holes. Tape the spool in place. Remove the straw.

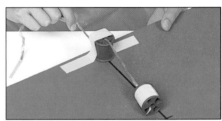

4 THREAD THE STRING attached to the first spool through the second spool and out the back of the board. Extend the string so that the spool reaches the 30 in (75 cm) mark. Tape the string at the back of the board to secure the pendulum.

5 PROP THE BOARD upright. Let the pendulum swing, but only through about 15°. If the swing is greater, the pendulum will not be so accurate. Time how long it takes to complete 20 swings. (Each swing crosses the center line twice.) Divide the time by 20 to find the time for one swing. Now time 20 swings with a pendulum of a different length. What is the difference? (Answer on p. 187.)

The pendulum crosses the center twice during each complete swing

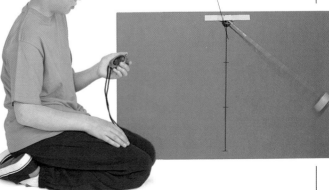

Speed

AN UNDERSTANDING OF SPEED is fundamental to modern life. Some of our aircraft and trains are designed to travel extremely fast, but speed limits are placed on roads used by ordinary cars, and radar traps catch drivers traveling too fast. Runners are timed so that they have a record of their speed and a clear target to beat. In all cases, speed is measured as distance traveled divided by time taken. Speed measurements are always relative. For example, the speed of an airplane can be given as "ground speed," relative to the ground below and not to the air speed of the wind around the aircraft. When we stand still, we would say that we are not moving. This is true relative to the surface of the Earth, but because the Earth rotates around its axis, and around the Sun, we are actually moving at gigantic speeds relative to the Earth's center or to other planets.

▓ The speed of sound

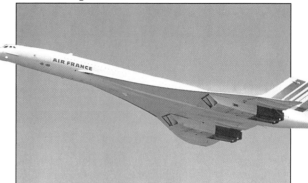

The speed of sound is about 760 mph (1220 km/h) at sea level. The first person to break the sound barrier in an aircraft was Chuck Yeager, a test pilot for the US air force, in 1947. At the moment, the Anglo-French aircraft Concorde is the only passenger aircraft able to fly supersonically. The speed of an aircraft relative to the speed of sound is known as the Mach number. An aircraft flying at Mach 2 is traveling at twice the speed of sound. Many fighter aircraft are now capable of traveling at high Mach numbers.

EXPERIMENT
Speed of rotation

Speed is not always defined in a linear way. Speed of rotation is given in revolutions per minute (rpm) or per second. For example, it is employed to monitor engines run on gas, which work best at particular rpm. This experiment shows the effect of changes in speed of rotation.

YOU WILL NEED
● string ● spool of thread ● adhesive tape

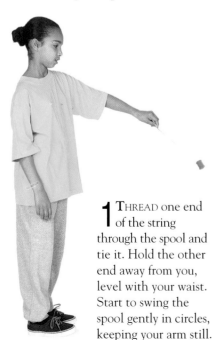

1 THREAD one end of the string through the spool and tie it. Hold the other end away from you, level with your waist. Start to swing the spool gently in circles, keeping your arm still.

2 SWING THE SPOOL as fast as you can. See how it rises as the speed of rotation increases. This effect is used to regulate the amount of fuel injected into the engines of some cars. The amount of fuel is then correct for the revolutions per minute (rpm) of the engine.

EXPERIMENT
Running speed

To find out how fast you can run, ask a friend to time you over a marked distance. Humans are not very fast over long distances. The fastest land animal is the cheetah, which can reach a speed of 60 mph (95 km/h) in short bursts.

Measuring running speed
To measure speed, divide the distance traveled by the time taken. Mark a distance of, say, 55 yd (50 m) and ask a friend to time you with a stopwatch as you sprint. Divide the distance by the time it took you in seconds to find the yards per second. To find your speed in yards per hour, multiply the speed figure by 3,600 (the number of seconds in an hour). To convert this new figure to miles per hour, divide it by 1,760.

EXPERIMENT
Wagon wheel effect

Images on film or television are transmitted at many frames per second. Our brain runs these together to make a moving picture. If a wheel spins on film or in front of a screen, and takes more than the time between two frames to turn through the angle between two neighbouring spokes, it may seem to turn backward.

YOU WILL NEED
● *ruler* ● *1 thick and 1 thin straw* ● *pencil* ● *pair of compasses* ● *protractor* ● *scissors* ● *poster board* ● *adhesive tape*

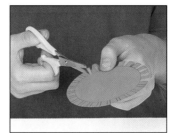

1 DRAW A circle with a radius of 2 in (5 cm). Mark its circumference at alternate 10° and 5° intervals, using a protractor. Draw another circle with a radius of 1½ in (4 cm), from the same center point.

2 MARK THE positions for 24 equally spaced slots on the wheel by drawing lines from the marked points on the outer circumference towards the center, stopping the lines at the inner circle.

3 USING SCISSORS, cut along all the lines, stopping at the inner circle. Cut out the smaller (5°) spaces to make slots on the edge of the wheel. Each slot should be about ¼ in (5 mm) wide.

4 SHORTEN THE thick straw to about 2 in (5 cm). Slash one end a couple of times so that it spreads out. Make a hole in the center of the circle and insert the straw so that the spread end forms a stopper.

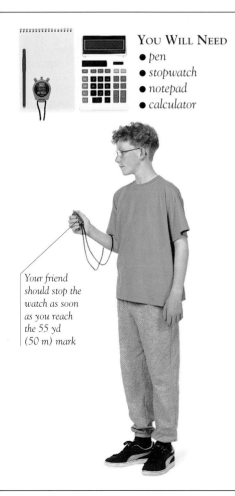

YOU WILL NEED
● *pen*
● *stopwatch*
● *notepad*
● *calculator*

Your friend should stop the watch as soon as you reach the 55 yd (50 m) mark

5 SECURE THE spread end of the straw to the circle with adhesive tape, leaving the central hole free.

6 INSERT THE thin straw through the center of the thicker one so that the wheel can be turned freely.

7 SIT IN FRONT of a television or a computer monitor, and turn it on. Hold the wheel level with your eye, in front of the screen. Twirl the wheel with your free hand and see the effect.

Twirl the wheel quickly so that it spins very fast

SHAPE

All around us

Everything has a shape, and the mathematics of shapes is called geometry. The geometrical patterns in a modern housing complex (left) show circular areas, straight lines, triangular roof construction, neat square or rectangular lawns, and a great deal of symmetry. Nature is equally ingenious in its symmetry and beauty, as in the identical designs on a butterfly's wings (above).

CIRCLES, RECTANGLES, spirals, and triangles were drawn by prehistoric peoples, whose art still exists. But these patterns existed in nature for millions of years before humans first evolved, in natural shapes. The understanding of geometry has been crucially important in the history of mathematics, providing a basis for spectacular theories about the motions of planets, perspective, and the double-helix spiral structure of the key component of life, DNA.

ANGLES AND SIMPLE SHAPES

Points, lines, angles, and planes make up the basis of all geometry. Regular angles and shapes are found throughout the natural world — for example, in honeycombs, crystals, and atoms. Studies have shown them to be particularly strong and efficient, whch is why we copy them in what we build, from huge bridges to satellites.

Geometric lines
Throughout history, humans have admired the aesthetic order in geometric shapes. Many ancient artifacts, such as this Greek vase dating from around 720 B.C., are adorned with geometric patterns.

Mathematicians call flat, straight-sided shapes "polygons." They describe a polygon by giving the length of each side and the angle (p. 116) between the sides. The simplest polygon is the triangle, which has three sides. If the sides are equal in length, the triangle is described as "equilateral." If two of the sides are of equal length, the triangle is described as "isosceles." All other triangles are "scalene."

The angle between two sides of a shape is usually measured in degrees (°). For example, a full turn is an angle of 360°, and the quarter-turn on the corner of a square or rectangle is an angle of 90°, also called a "right angle" (p. 124). The circle is divided into 360° for historical reasons: this was the number of days that made up the ancient Babylonian year.

Pyramids
The Pyramids of ancient Egypt, built more than 4,000 years ago, are still among the most striking examples of structures that incorporate triangles. These enormous stone buildings consist of sloping triangular walls arranged around square bases. The tallest

of the Pyramids, the great Pyramid of Khufu, was originally 482 ft (147 m) high.

The ancient Egyptians found that if a pyramid is cut vertically along the diagonals of its base, four smaller, identical solids will be produced. Each of these solids has a cross-section that is a right-angled triangle (a triangle with an angle of 90°). In the Rhind Papyrus (p. 14), Ahmes used this discovery to show how the height of a pyramid is related to the size and angle of slope of each triangular wall. He produced a table of ratios to which pyramid builders could refer to make sure that all four sides of a pyramid sloped by the required angle. Ahmes' table was an early forerunner of tables of special ratios, called "trigonometric" ratios, which mathematicians still use today.

Pencil case
This pencil case has been made from a parallelogram twisted to form a cylinder. The length of the parallelogram's short sides is equal to the circumference of the cylinder's end.

Greek theories
The ancient Egyptians' understanding of shapes and angles was based on ideas they had discovered through practical experiments. In contrast, the ancient Greeks developed their ideas on shapes and angles using carefully crafted theories and mathematical proofs. The ancient Greek philosopher Pythagoras (p. 124) knew many important properties of triangles. One of these was that the sum of the

three internal angles in a triangle is 180° (one-half of a full circle). In addition to studying the properties of individual shapes,

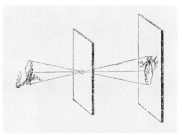

Camera obscura
When light rays are concentrated on a lens or on a small point, as in this camera obscura, they have to cross in order to pass through it. Like any other lines that cross, they form two isosceles triangles whose sides slope at the same angle. Modern cameras, and animals' eyes, collect light in the same way.

Pythagoras was also interested in the way shapes fitted together. He found that he could fit equilateral triangles, squares, or hexagons (six-sided flat shapes) like tiles on a floor, without any gaps between them. We now say that shapes with this property "tessellate" (p. 130), after the Latin word *tessera*, meaning a small tile or piece of mosaic. Tessellation is one form of a property called symmetry (p. 158).

Euclid's geometry
The ancient Greek mathematician Euclid (*c.* 330–275 B.C.) made investigations of shapes and angles. In his book *Elements*, he wrote many elegant proofs of well-known theories. Euclid's work was so thorough that we now refer to all geometry of lines, points, shapes, and solids as "Euclidean geometry."

Far from being an abstract mathematician, Euclid showed many ways in which his work could be applied to problems in the real world. For example, he discussed why some shapes, such as triangles, are rigid (p. 123) while others, such as squares and rectangles, are not. The biggest structures today, such as skyscrapers and bridges, are built using these principles. Euclid also suggested that geometry could be employed to trace the paths of light rays. Geometry is still used in this way in optics today.

All Euclidean geometry relies on five facts called "postulates" that we come across every day. The most important of these postulates concerns parallel lines (lines that never meet each other, no matter how far they are extended).

The system of geometry that was outlined by Pythagoras, Euclid, and their Greek successors has been used for centuries in mathematics and science. It is only during the last 150 years that mathematicians have been able to develop totally new ideas, which have extended mathematical knowledge beyond the principles first laid down by these Greek scholars in ancient times.

Ratios in triangles

One of the most important branches of geometry the ancient Greeks developed is trigonometry. This subject depends on particular properties of triangles and other shapes. If a triangle (or any other shape) is increased in size by multiplying the lengths of all its sides by the same amount, all of the angles within the triangle will remain unchanged.

After the ancient Egyptians, one of the first people to develop trigonometry was the ancient Greek astronomer Hipparchus (p. 126). He recorded the locations of no fewer than 1,080 fixed stars, and worked out the motion of the Moon relative to the Earth. He developed an early form of trigonometry to help him to work out the paths of heavenly bodies. Hipparchus used this idea to produce a set of ratios that could be used to estimate the lengths of the sides of a triangle from its angles. The ratios with which he worked are now known as "trigonometric ratios."

Hipparchus' work was rewritten by Ptolemy and later used by Arab scholars such as Albuzjani (A.D. 940–998). Like the Greeks, the Arabs used trigonometry to trace the paths of the Moon, the planets, and the stars. Albuzjani wrote a new set of ratios very similar to one set of ratios, now known as tangents (p. 127), that we use today. Trigonometry is now employed, indirectly, to measure distances, such as the heights of mountains or buildings. It has also been used to study microscopic structures. One of the most crucial modern applications is in crystallography, the study of crystals. Substances that exist as crystals, such as salt and sugars, are formed from atoms arranged

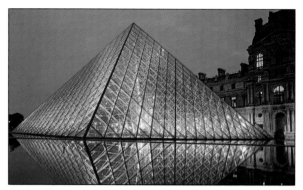

Louvre pyramid
Straight lines projected upward from the corners of a square and meeting at a central point form a pyramid such as this glass structure, which stands outside the Louvre in Paris, France.

in a regular structure called a "lattice." This regular structure is what gives crystals their consistent shape.

In 1913, the German physicist Max von Laue (1879–1960) found that X rays were scattered when passed through crystals. Soon afterward, a British physicist, Lawrence Bragg (1890–1971), and his father, used trigonometry to show how the structure of crystals could be estimated by calculating the angle at which the crystals scattered X rays.

Crystallography can be used to study other substances as well as crystals. In 1938, a medical researcher, Rosalind Franklin (p. 149), used Bragg's ideas about X-ray diffraction to uncover the structure of DNA. Her studies showed that DNA had to be in the shape of a helix. This work led to the formation, by Crick and Watson, of their model showing the DNA molecule as two helices wound together.

Tessellating shapes
Some flat shapes fit together exactly. This feature can be used to make patterns, as with these rhombi and triangles (left) and octagons and squares (right).

Mirrors in parallel
Light can travel over long distances or around bends by repeated reflection. These mirrors all reflect light at the same angle. The torch must be held at this angle so that the light hits all of the mirrors.

Light travels in straight lines from one mirror to the next

What is an angle?

AN ANGLE IS A WAY OF EXPRESSING the amount by which a line or other object turns. It can be formed by two straight lines meeting at a point, or in other ways, such as by a vehicle traveling around a corner. Angles are usually expressed in degrees, by means of the symbol °, and are classified according to their size. A full turn — when an object rotates about one end point and returns to its original position — is 360°. A right angle is 90°; an angle less than 90° is an acute angle; one between 90° and 180° is obtuse; and one that is more than 180° is a reflex angle. In advanced mathematics, such as calculus (p. 71), angles are measured in radians, written with the abbreviation "rad." The Greek astronomer Hipparchus (p. 126) is credited with dividing a circle into 360°. He inherited the idea from early astronomers, who believed that the Earth was stationary and stars revolved around it on a circular band of 12 parts, each comprising 30 days, roughly equal to one cycle of the Moon.

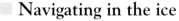
Navigating in the ice

Navigators take bearings in degrees, measured clockwise from magnetic or true north. This picture shows a path being cut through ice for ships on a 19th-century Arctic expedition. Compass readings would have been vital in planning the ships' course.

EXPERIMENT
Finding true north

The Earth has a magnetic field that affects a compass. What a compass shows is magnetic north, which is different from true north. This difference is called variation or declination. The magnetic north pole is in northern Canada, about 1,000 miles (1,600 km) from the North Pole. To find true north, you need to go outdoors early in the morning on a sunny day and take readings between 0900 and 1700 hours. Shadows are shortest at noon, when the Sun is at its zenith (maximum angle in the sky). In the Northern Hemisphere, shadows point directly north at noon, whereas in the Southern Hemisphere they lie at 180° from true north.

YOU WILL NEED
● *ruler* ● *8 in (20 cm) long dowel* ● *pens* ● *pencil* ● *scissors* ● *thumbtack* ● *thread* ● *compass* ● *large sheet of poster board*

Rule a line from the center along the length of the shadow

1 INSERT THE THUMBTACK in the center of the poster board, and tie a length of thread to it. Tie a pencil 10 in (25 cm) down the thread, and use it to draw a circle. Cut out the circle and place it on the ground. Make a hole in the center, and push the dowel through it, to fix the circle to the ground. Every hour, mark the length of the shadow cast by the dowel.

2 LAY THE COMPASS at the center of the circle, and align the N with the shortest shadow line. Mark where the needle points (magnetic north), and rule a line from the compass to this mark. The angle created shows the difference between true and magnetic north.

Arrow points to true north

Compass needle points to magnetic north

Shadow line

EXPERIMENT
Making a spirit level

A spirit level is an indispensable tool for carpenters and builders. It is laid on a horizontal surface, and the position of the bubble inside will indicate whether or not the surface is level. If it is, the bubble will lie exactly in the center. This simple spirit level measures only very small angles in relation to horizontal surfaces.

YOU WILL NEED
- *strip of wood* ● *scissors*
- *pen* ● *modeling clay* ● *adhesive tape* ● *food coloring* ● *parcel tape*
- *clear plastic tube* ● *funnel*
- *jug of water*

1 CUT A 12-IN (30-CM) length of tubing. Make sure the inside and outside are clean, then seal one end with modeling clay.

2 POUR COLORED WATER carefully into the plastic tube until it is almost full. Make sure you leave a little space at the end so an air bubble can form.

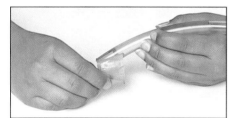

3 SEAL IN THE WATER with another small piece of modeling clay, leaving a gap for the air bubble. Wrap plenty of parcel tape neatly around both ends of the tube. This prevents the water from leaking out and limits the loss of water through evaporation.

4 ATTACH THE water-filled tube securely to the strip of wood with adhesive tape. If you first position a piece of modeling clay under the center of the tube, so that the tube bends upward slightly, the spirit level will give more accurate readings.

5 LAY THE SPIRIT LEVEL on a flat surface such as a table or a shelf. Wait until the air bubble comes to rest roughly in the middle of the tube, and mark the ends and the center of the bubble with an ink pen. This will be your reference for a totally horizontal, flat surface.

INVESTIGATING SHAPES
Classical mathematics

The Greeks believed that there were five basics from which geometry could be demonstrated.

RULE 1
A circle is bisected (p. 182) by its diameter.

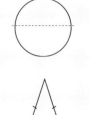

RULE 2
In an isosceles triangle, with two equal sides, the two angles opposite these sides are equal.

RULE 3
When two straight lines cross, the opposite angles at the crossing are equal.

RULE 4
Any angle inscribed in a semicircle, so that a triangle is formed with the diameter, is 90°.

RULE 5
Triangles with all three sides the same length are congruent (identical).

Using the spirit level
This primitive spirit level is best used as a rough guide in checking that a surface is horizontal. When the spirit level is placed on a surface and the bubble lies exactly between the two marks, you will know the surface is level.

Secure the tube at several points to keep it steady

The lines show the bubble's position when the spirit level is on a flat surface

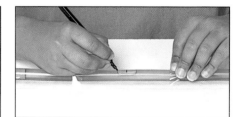

Using the angle

THE ANCIENT GREEKS STUDIED ANGLES and invented many rules concerning them. They also discovered rules that govern how angles in triangles and circles are related. These mathematical proofs are still used today by engineers, navigators, designers, and surveyors. Early astronomers used astrolabes (p. 97) to measure the angle of elevation of the stars, and by that means calculated the distance to the stars and the circumference of the Earth (p. 135). Trigonometry (pp. 126–127) is based entirely on angular measurement and the relationship of an angle to the sides of a triangle.

Angles in architecture

Angles play an important role in architecture (p. 126). The French architect Le Corbusier (1887–1969) completed the Chapelle de Notre Dame du Haut, a place of pilgrimage in Ronchamp, France, in 1955.

EXPERIMENT
Defying gravity

We all know that you cannot roll something uphill without some force to push it. The roller in this experiment is made from the tops of two plastic bottles. It appears to defy the law of gravity and to roll uphill by itself. The secret lies in the angle of the slope and the shape of the bottle tops. If you watch the bottom of the roller closely as it rolls, you will see that it is in fact falling. You need to think clearly about this experiment, or the illusion will cause you to believe that the bottle is defying gravity.

YOU WILL NEED
- *2 lengths of dowel* ● *2 plastic bottles* ● *modeling clay* ● *adhesive tape* ● *scissors* ● *books*

1 USE THE SCISSORS to cut off the top sections of the two plastic bottles about where the straight parts of the bottles begin. Cut carefully, leaving no jagged edges. Make sure the bottle tops are the same depth — in other words, that they are identical in size.

2 ABUT THE TWO SECTIONS together at the cut edges. Secure them together firmly, putting strips of tape across the joint and then around the joint.

Wood held firm with modeling clay

Getting the angle right
The dowels should be arranged so that the pieces are closer together at the bottom end of the slope. You will have to experiment to get the angle right.

3 ARRANGE SOME BOOKS in two piles — one pile with two or three more books than the other. Use the illustration below left as a guide. Use equal-sized pieces of modeling clay to create a slope with the two dowels.

4 REST the bottle roller on the lower end of the slope and see what happens.

Center the roller between the two dowels

EXPERIMENT
Leaning tower

Most people have seen pictures of the Leaning Tower of Pisa in Italy and wondered why it does not fall over. Work began on the Tower in 1174 but construction problems — the foundations settled unevenly — gave it its distinct lean. The angle that something can lean away from the vertical before it falls over is determined by the position of its center of gravity and the size of its base. As the angle of lean increases, the center of gravity moves in relation to the base and eventually, when the center of gravity acts downward outside the base, the object will fall over.

YOU WILL NEED

● plastic bottle with lid ● pencil ● pen ● hole puncher ● pair of compasses ● poster board ● foamcore ● food coloring ● thumbtack ● modeling clay ● scissors ● adhesive tape ● protractor ● ruler ● wooden board ● jug of water ● books

1 To make a large protractor, draw a semicircle on the foamcore. Using an ordinary protractor as a template, mark the angles at 10-degree intervals. Rule the lines for the angles.

2 Cut out the semicircle of the protractor. Next, draw a thin strip of poster board, about one-quarter the length of the plastic bottle and about ½ in (15 mm) wide. Cut out the strip.

3 Add food coloring to the water. Use a generous amount, to create a strong color. Then pour the water carefully into the plastic bottle until the bottle is two-thirds full.

5 Stick the thumbtack through the top of the poster board strip and secure to the bottle by sticking it into the modeling clay. Do not push too far or you will puncture the plastic bottle.

The pin holds the strip securely but also allows it to swing

4 Using adhesive tape, secure a lump of modeling clay halfway between the base of the bottle and the level of the water. This is the approximate position of the bottle's center of gravity.

The gravity indicator is important in this experiment

6 Rest one end of the board on some books to make a slope. With modeling clay, secure the protractor to the table at the bottom of the slope so you can read off the angle of incline. Stand the bottle on the slope. The card strip will hang vertically, showing the direction of gravity. Increase the angle of the incline with more books, or by tilting the board with your hand, and see at what point the bottle falls over.

Hand-made protractor is used to read off the angle of lean

Straight lines

THE SIMPLEST SHAPE in geometry is a straight line. A line has only one dimension — length. Straight lines are found everywhere. In nature, light travels in a straight line and objects that are set in motion will move in a straight line unless some other force intervenes. A straight line drawn between two points is the shortest way of getting from one point to another. Parallel lines are straight or curved lines (such as the edges of the Möbius strip, p. 165) that stay the same distance from each other, no matter how far they are extended.

III Keeping in touch

Overhead wires provide the current that makes electric streetcars run, and they must remain in contact with the streetcars. Connectors based on flexible, straight-sided shapes are used because they can be compressed and extended depending on the evenness of the streetcar's route.

EXPERIMENT
Parallelogram pencil box

A parallelogram has four straight sides that form two pairs of parallel lines, and has two pairs of equal angles. This experiment, using the inner tube of a kitchen-paper or toilet-paper roll, shows how the shape can make a strong cylinder. A tube made from a rectangle would not be so strong; the seam of the join (a point of weakness) would run in one direction, instead of curving evenly around the tube, so making the shape prone to collapse.

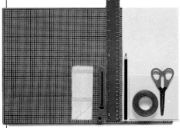

YOU WILL NEED
● *toilet roll inner tube* ● *craft knife* ● *ruler* ● *pencil* ● *adhesive tape* ● *scissors* ● *poster board* ● *foamcore* ● *cutting mat*

Adult help is advised for this experiment

1 LAY THE opened toilet roll on the poster board. Using a ruler, draw around the shape. Cut it out.

2 FIX adhesive tape to one long side of the shape. Leave half the tape's width free of the poster board.

3 TWIST THE SHAPE so that the ends align, and stick the cylinder shape together along the long sides.

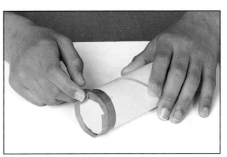

4 STAND the tube on the foamcore, and draw around the circular end. Ask an adult to cut out the circle with the craft knife.

5 TAPE THE CIRCLE onto one end of the tube. Stick half the tape's width to the tube. Make vertical cuts in the loose tape and fold it onto the circle.

The tube, strengthened with tape around the top edge, makes an attractive pencil box

Types of lever

One of the simplest machines is a lever — a straight object that uses a fulcrum, or balancing point, to lift or move a load. Some levers enable a small force, applied at one point on the lever and moved through a large distance, to move a larger weight or force, at another point, through a smaller distance. Other types transform a small movement into a larger one. Chopsticks can work as levers. The lower stick is kept still and the top one is moved. The levers turn the movements of the user's fingers into smaller ones with greater force. This allows the user to maneuver the ends of the sticks with tiny, exact movements.

Fulcrum – the point around which the levers move

Puzzle

When a rectangle is sheared — pushed over to one side — it is difficult to estimate its size accurately. Collect together a number of coins of different sizes. Estimating the size by eye only, see if you can select a coin that fits inside these two shapes exactly. What do you notice about the coins that fit each shape? (Answer on p. 187.)

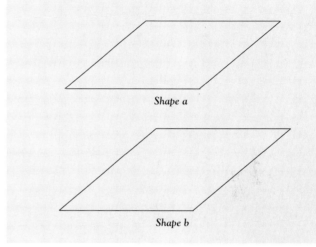

Shape a

Shape b

EXPERIMENT
Bending light

Light usually travels in straight lines. When light is reflected, for example in a mirror, the angle of incidence (angle at which the light hits the mirror) is equal to the angle of reflection. In this experiment, the mirrors are placed so that the light is reflected from one mirror on to the next, and so on. Periscopes in submarines, and cardboard ones for viewing over the heads of crowds, also use mirrors to bend light.

YOU WILL NEED
● *flashlight* ● *3 small mirrors* ● *piece of black card the same size as the mirrors* ● *modeling clay*

1 ARRANGE THE MIRRORS in two parallel lines, as shown on the right of the page, and secure them to the tabletop with modeling clay. Put the black card at the end of one line of mirrors.

2 DARKEN THE ROOM. Lay the flashlight on the table so that it points toward the mirror at the opposite end from the card. Make sure the angle it makes with the first mirror allows the light to be reflected from that mirror to the next. Switch on the flashlight to see how the beam of light behaves.

See if the light from the flashlight is reflected in all the mirrors

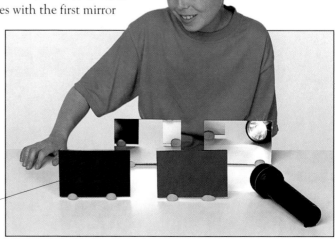

Triangles

A TRIANGLE IS A SHAPE with three straight sides and three angles. It has the minimum number of straight lines needed to make a closed shape. The most regular form is the equilateral triangle, with sides of equal length and three equal angles, each of 60°. Other forms include the isosceles triangle, with two equal sides and angles; the right-angled triangle (p. 124), which has one angle of 90°; and the scalene triangle, with no equal sides or angles. The shape is crucial to engineering and architecture, to make rigid, strong structures such as rafters for the roof of a building, and constructions such as bridges and geodesic domes. In surveying and navigation, a technique called triangulation is used to find the distance of a given point from two other points a known distance apart, by considering all three as forming a triangle and measuring the angles between them. This is necessary whenever it is not possible to calculate distance in one step — when measuring around corners, for instance, or digging a tunnel.

Triangles in structures

Look at almost any man-made structure and you will find groups of triangles. Pitched roofs, for example, are used around the world, particularly in places where there is heavy rain or snowfall. The triangles of wood in the roof transform large forces that act on each triangle's topmost vertex (p. 185) into two smaller forces, which act along the sides of the triangle that meet at that vertex. The roof load can then be supported equally by the walls.

Puzzle

How do you cross a 10 ft (3.05 m) wide castle moat, with quarter-circle corners, if you have two 9 ft (2.74 m) planks? (Answer on p. 187.)

EXPERIMENT

Finding the sum of the angles of a triangle

Although there are various types of triangle and they look very different, they all share some properties. Try the experiment below with several forms of triangle (such as isosceles and scalene), compare the results, and see what you discover about the sum of the angles within each shape. Is there a difference with each triangle, or does some general feature hold true for all of them? How many degrees are there within each triangle?

YOU WILL NEED
- *ruler*
- *pen*
- *scissors*
- *protractor*
- *paper*

1 DRAW A TRIANGLE. Make the shape fairly large. Mark all the angles; draw squares for right angles and curved lines for other angles. Shown here is a scalene triangle.

2 CUT OUT the paper triangle, then cut off all the corners. Make the cut pieces reasonably large, so you can measure them easily with the protractor.

3 LAY THE CUT PIECES on top of the protractor, with the edges touching and the angles next to each other. The points of the angles should touch the center of the protractor. What do you notice about the three angles? (Answer on p. 187.)

🧩 Puzzle

Look carefully at the squares in the picture below. How many triangles can you see in the large square? As you count the triangles, it may help if you look at the square, section by section, and make a note of the number of triangles in each section, then add the numbers together. (Answer on p. 187.)

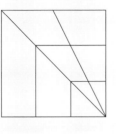

(Answer on p. 187.)

EXPERIMENT
Making a pinhole camera

Cameras and your eyes make use of triangles formed by light from images. Use this "camera" (and perhaps cover your head to block out light around the screen) to see how images are received on a lens.

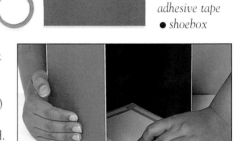

YOU WILL NEED
- *ruler* • *aluminum foil* • *tracing paper* • *colored and flat black paper* • *glue* • *pencil* • *pin* • *pair of compasses*
- *scissors*
- *double-sided adhesive tape*
- *shoebox*

1 COVER THE BOX with black paper inside and colored paper outside. Cut a hole in one end; leave a 1-in (2.5-cm) frame. Cut a circle 1¼ in (3 cm) across in the other end. Cut tracing paper to fit over the frame, and fix it with tape.

2 CUT OUT A 3-IN (8-cm) square of foil. Tape the foil over the round hole and make a tiny hole in the center with a pin. This is your lens.

3 STAND in a dark place, and point the pinhole at a window or well-lit area. You should see an upside-down image on the tracing paper.

How the camera works
Light is reflected off the object you see in many directions. Some light from the object travels through the tiny hole and hits the screen at the back of the box. Because the straight light beams have to cross in order to go through the hole — as they do when light passes through the lenses in our eyes — the final image is upside down.

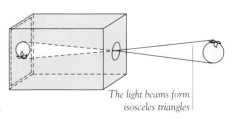

The light beams form isosceles triangles

EXPERIMENT
Strength in triangles

Triangles are used in man-made structures for a good reason: they can form a rigid shape even when made from flexible or weak materials. Make a triangle by joining drinking straws together, following the steps below, and compare it to a square or another shape made in the same way. What do you notice about the strength and rigidity of each shape that you have made? (Answer on p. 187.)

(Answer on p. 187.)

YOU WILL NEED
- *drinking straws*
- *push pins*

1 HOLD TWO DRINKING STRAWS together with their ends aligned. Stick a pin through both straws, just below the ends, to join them together. Open the straws to form a "V."

2 ADD another straw and fasten the ends to the first two straws. When you have finished constructing your triangle, try to bend it out of shape. What do you notice?

Right angles

THE RIGHT ANGLE, measuring 90°, is probably the most important angle used in geometry, science, and engineering. Graphs in two or three dimensions have axes at right angles to each other (p. 74). The vertical is at 90° to the horizontal, gravity acts at 90° to the horizontal, and the tangent of any circle is at 90° to its radius. The main points of the compass (N, S, E, and W) are all at right angles to one another. The right-angled triangle, with one angle of 90°, is similarly important. For example, it is the basis of trigonometry (pp. 126–127). This form of triangle was certainly used by the Egyptians when they built the pyramids. There is even evidence that they used Pythagoras's theorem (below), although they did not write it down.

EXPERIMENT
A 3:4:5 triangle

Pythagoras' theorem states that in a right-angled triangle the square of the hypotenuse (the longest side) is equal to the sum of the squares on the other two sides. You can also discover that any triangle in which the sides are in the ratio of 3:4:5 (p. 54) will always be a right-angled triangle; $3^2+4^2=5^2$. This fact was used by builders in ancient times to construct accurate right angles.

YOU WILL NEED
● *foamcore*
● *3 thumbtacks*
● *scissors* ● *string*
● *adhesive tape*

1 CUT A PIECE of string about 36 in (90 cm) long. Using your index finger as a unit of measurement, mark 12 equal lengths along the string. Mark each finger length with a small tab of adhesive tape.

2 LAY the string on the foamcore. Shape it into a triangle, with four units across the bottom and three up the side. Place a thumbtack at each corner to keep the string taut. You will see that you have created a right-angled triangle.

The Pythagorean school

Pythagoras (*c.* 582–500 B.C.) was a prophet, mystic, and mathematician. He founded a school at Crotona, Italy, which became a secret society. The scholars there were called Pythagoreans, and their beliefs spread throughout the Greek world. Their motto was "All is number," and they proved the fundamental theorems of plane and solid geometry. One scholar, Philolaus (who lived in the late 5th century B.C.), put forward a theory about

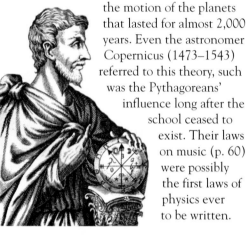

the motion of the planets that lasted for almost 2,000 years. Even the astronomer Copernicus (1473–1543) referred to this theory, such was the Pythagoreans' influence long after the school ceased to exist. Their laws on music (p. 60) were possibly the first laws of physics ever to be written.

Avoiding detection

HMS *Monmouth* is a British Royal Navy frigate. It uses the latest technology to minimize detection by radar, infra-red, and magnetic sources. Radar can detect objects by measuring the time radio waves take to travel to the object, be reflected, and return. The more vertical surfaces there are on the object, the more it reflects radio waves back toward the radar. The ease of a ship's detection by radar is called its "signature." The *Monmouth* has a low signature because it has few right angles; all its vertical surfaces are built at a skew of 7°.

EXPERIMENT
Making movement with electric current

This simple motor, which converts electricity into motion, is not very different from that made by Michael Faraday (p. 29) in 1821. He used the dangerous substance mercury, but this experiment works with another conductor — salt water. Faraday showed that an electric current interacting with a magnetic field could produce movement at right angles to both of them. The wires and the water all conduct the electric current. The direction of the current and the movement can be worked out using right angles (see below).

YOU WILL NEED
- *aluminum foil* - *poster board* - *strong, small magnets* - *copper wire* - *4.5-volt battery* - *aluminum pie dish* - *modeling clay* - *battery wire connections* - *metal coat hanger* - *warm salt water* - *ruler*

Adult help is advised for this experiment

1 LAY A SHEET OF ALUMINUM foil on the tabletop. Put a small square of poster board against one edge. Roll the lump of modeling clay into a roughly cylindrical shape, and put it on top of the board.

2 CUT A PIECE from the coat hanger. Bend it into a long "L" with a hook on the short end. Stick the other end into the clay. The poster board prevents the metal from touching the foil.

3 PUT THE MAGNET in the center of the pie dish. Measure the height of the wire hook from the foil, and cut a piece of copper wire to this length. Place the dish under the hook.

4 STRAIGHTEN the copper wire, and make a small hook at one end. Pour warm salt water into the pie dish to just below the rim. The warmer and saltier the water, the better your results will be.

5 HOOK THE COPPER WIRE onto the coat hanger wire and let it rest in the water. Attach the positive battery terminal to the foil and the negative one to the L-shape. The end of the copper wire should start to swing around the magnet, and the salt water will fizz as hydrogen is given off. After this "motor" has been running for a while, unplug the battery. Empty the dish, and you will see tiny holes in the base where the aluminum has dissolved.

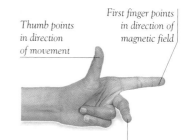

Thumb points in direction of movement

First finger points in direction of magnetic field

Second finger points in direction of current

Fleming's left-hand rule
This rule, devised by the English physicist John Fleming, is a way of remembering the direction of the movement, current, and magnetic field in electric motors. The current acts in one direction. At right angles to this is the motion of the wire turning around the magnet. And at right angles again is the magnetic field.

Trigonometry

TRIGONOMETRY IS BASED ON THE STUDY of right-angled triangles (pp. 124–125), a form of geometry that developed out of the study of the stars. Theorems on the ratios of the sides of these triangles had been used by the Egyptians and Babylonians, but the ratios that we use today were first set out in about 150 B.C. by Hipparchus (right), who arranged them in tables. François Viète (p. 68), in the late 16th century, applied trigonometry to number theory and algebra as well as to geometry, even using it to solve a cubic equation (p. 72). Trigonometry has many practical applications, for example in building, civil engineering, and navigation. It is used in situations where measurements cannot be made physically — for instance, in finding the distance to a star or to an island. Angles in triangles and other shapes have many characteristics that remain true for angles of the same size no matter where they occur, most notably their trigonometric functions, such as sine, cosine, and tangent (see opposite page). These characteristics are extremely important to engineers and surveyors.

Hipparchus

Hipparchus of Nicea (c. 170–125 B.C.) was fascinated by astronomy and he used mathematics in his astronomical studies. His work was the earliest attempt at devising formulas for trigonometry. He applied mathematics to define the positions of cities on the Earth's surface by using lines of longitude and latitude, a system that is still employed today.

Mecca

Mecca, in western Saudi Arabia, was the birthplace of the prophet Muhammad, the founder of Islam, around A.D. 570. It is the holiest city in Islam. Muslims pray toward it five times daily, and millions also make a pilgrimage to the city at least once in their lifetime. Here people pray at the Kaaba, a shrine in Mecca. For centuries, Muslims used trigonometry and the stars to navigate across the desert to Mecca.

How high is the mountain?

The heights of mountains used to be found by classical surveying techniques based on simple trigonometry. A surveyor would use a theodolite and a tape measure and information about right-angled triangles. Today, surveying tools use infra-red reflectors, and information from satellites, to get an accurate basis for readings.

Modern theodolite
A theodolite is an instrument with a sight lens that is used to measure horizontal and vertical angles. The instrument shown here finds the angles using a beam of infra-red light. It also has a memory unit to store the data gathered during use.

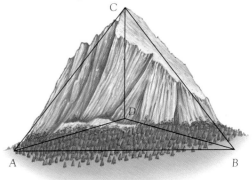

Using angles to find the height of a mountain
The base and height are plotted as an irregular tetrahedron (p. 152), made of three right-angled triangles, and another triangle ABC, where C is the summit. D is directly beneath the summit, at the same altitude as A and B. The surveyor measures the baseline AB, between two points at the same known altitude, then finds the angles CAB, CBA, CAD, and CBD. The angles and known lengths are used to work out CD, the height of the mountain above the baseline.

INVESTIGATING SHAPES
Trigonometric functions

In trigonometry the ratio between any two sides of a right-angled triangle is given as a function (p. 73) of angles within the triangle. These ratios (p. 54) are called trigonometric functions. The three most commonly used ratios are sine, cosine, and tangent. They can be applied to find unknown lengths of sides or unknown angles. Scientific calculators work out these ratios for you (p. 19).

The angle
In diagrams and calculations, the angle to be measured is usually shown using the Greek letter θ (theta).

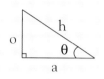

The sine
For an angle θ, the sine (usually abbreviated to sin) is the ratio of o, which is the length of the side opposite θ, to h, which is the length of the hypotenuse.

$$\sin θ = o/h$$

The cosine
For an angle θ, the cosine (cos) is the ratio of a, the length of the side adjacent to θ, to h, the length of the hypotenuse.

$$\cos θ = a/h$$

The tangent
For θ, the tangent (tan) is the ratio of o, the length of the side opposite θ, to a, the length of the side adjacent to θ.

$$\tan θ = o/a$$

EXPERIMENT
Finding the height

Trigonometric ratios remain constant for a given angle no matter what the size of the triangle. Knowing this, you can find the height of an object like a house or tree without using any special instruments, simply by looking through your legs. By creating a right-angled triangle in which a friend and the ground form two sides of the triangle, you are getting information that will help you to find the height of something that is much taller. The equation to use when you have gathered all the necessary information is: Height of tree = Your friend's height divided by the number of steps she is away from you, multiplied by the number of steps you are away from the tree when you bend down.

YOU WILL NEED
- *calculator* ● *notepad*
- *pencil*

The line of sight between your eyes and the top of your friend's head is the hypotenuse

1 WALK AWAY FROM YOUR FRIEND until you can just see the top of her head when you bend down and look through your legs. Make a note of how many steps you have taken, and of your friend's height. Assuming you always bend to look through your legs in the same way, then you have defined the tangent of the angle in this particular triangle (see definition above). Because the tangent will always be the same for this angle, you can now use the formula to find the height of something taller, perhaps a house or a tree.

When you bend down your head is closer to the ground and forms the angle to be used

Your friend is one side of the triangle

The point on which your friend stands is the right angle of the triangle

2 WALK AWAY FROM THE TREE, counting the number of steps you take. Bend to look through your legs in the same way as before. When you can see the top of the tree, and no higher, make a note of the number of steps you have taken. The height of the tree can be found by the equation in the introduction above, because the tangent of the angle remains the same.

The triangle formed with the tree is much larger than the one formed with your friend, but still has the same angles

Simple shapes

FLAT, OR PLANE, SHAPES with three or more sides are called polygons. The simplest are triangles, with three sides, and quadrilaterals, with four. The most common quadrilaterals are squares, rectangles, and parallelograms. Polygons with sides of equal length and equal interior angles are known as "regular." Plato and Euclid first analyzed these shapes almost 2,500 years ago. Their work is the basis of much more complex analyses in modern science and computing. Regular polygons often occur naturally. Examples include molecules, the units of any chemical element. Benzene, for instance, has a hexagonal structure. Biologists refer to polygons when studying viruses, because many have regular forms, such as the icosahedron (p. 152), a shape made of 20 equilateral triangles.

Geometrical beehive

Honeycombs are used to store honey and house larvae. The cells in the comb are made of wax, with a regular hexagonal cross-section. They fit together perfectly, to prevent dirt or predators from entering. Squares or triangles would fit equally well, but hexagons provide the largest storage volume for the amount of wax used.

EXPERIMENT

Magic wallet

This magic wallet is made of two rectangles joined with ingenious paper hinges. It can be opened from either edge and is used to hold paper money — with one action it can secure or release the money. To make this wallet you will need to follow the template below.

YOU WILL NEED
● *ruler* ● *paper* ● *thick poster board* ● *pencil* ● *glue* ● *scissors*

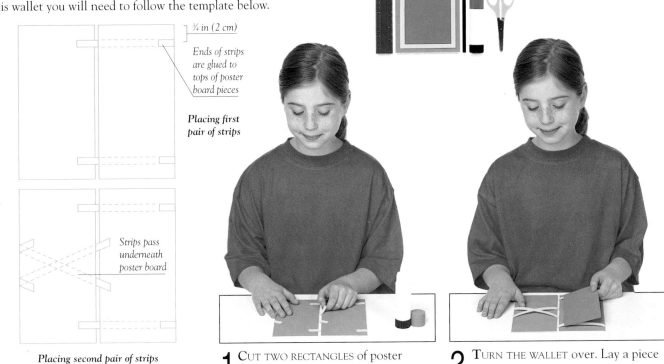

¾ in (2 cm)

Ends of strips are glued to tops of poster board pieces

Placing first pair of strips

Strips pass underneath poster board

Placing second pair of strips

Attaching the paper hinges
Lay the strips under the poster board pieces so that the ends are visible. Glue the ends to the tops of the boards so that the rectangles are joined at the middle.

1 CUT TWO RECTANGLES of poster board, each 7×3½ in (18×9 cm). Then cut four strips of paper, each 5½×½ in (14×1.2 cm), for the hinges. Attach the hinges as shown. Glue down the ends.

2 TURN THE WALLET over. Lay a piece of paper over the horizontal strips on the left-hand side. Close the other side over it. See what happens when you open the wallet again at the right-hand side.

EXPERIMENT
Angles in a quadrilateral

The word *quadrilateral* means "four-sided." Quadrilateral shapes include squares, rectangles, rhombi, kites, and parallelograms. Some of these shapes have sides of different lengths, some have slanted edges, but they all share one particular feature concerning the internal angles at the vertices. Try this experiment with different quadrilaterals, and see if you can find out their common characteristic. (Answer on p. 187.)

YOU WILL NEED
- *ruler*
- *pen*
- *scissors*
- *poster board*

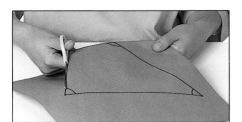 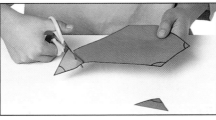 

1 DRAW A QUADRILATERAL of any shape you want. Shown here is an irregular quadrilateral that has no equal sides or equal angles. Cut out the shape.

2 MARK ALL THE ANGLES. Draw small squares on right angles and curved lines on other angles. Cut off the angles, making the cut pieces reasonably large.

3 LAY THE PIECES on a table with the angles touching. What is the sum of these angles? Is it different from the sum of the angles of a triangle (p. 122)?

EXPERIMENT
Pentominoes

Pentominoes are similar to dominoes except that each piece is made up of five connecting squares rather than two. There are 12 possible shapes. In this game the pentominoes are fitted together to cover a chessboard with four squares left uncovered. It is estimated that there are 1,000 ways to fit all the pieces on the board. Of these, 65 have the left-over squares exactly in the center. In this game, which is for two players, the loser is the first player who is unable to fit any of the pentominoes on the board.

Template for pentomino pieces
Use this template as a guide when drawing the 12 shapes. Outline the pentomino edges in pen. Make each shape a different color. The shaded squares are not part of any of the shapes.

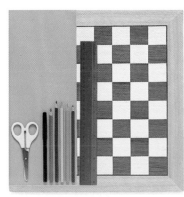

YOU WILL NEED
- *scissors* • *pen*
- *colored pencils* • *poster board* • *ruler* • *chessboard*

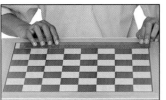 

1 MEASURE the squares on your chessboard. Draw a grid of 8×8 squares on the poster board, using squares the same size as those on the board.

2 DRAW and color the 12 pieces, following the template above. Cut them out. (The four left-over squares can be made into a larger square.)

3 PUT THE PENTOMINOES in a pile between you and a friend, and take it in turns laying the shapes on the chessboard. The pieces can be laid either right-side up or upside down. The first player who cannot move is the loser. With practice, it is possible to plan ahead and force the other player to lose.

Fitting shapes

Wʜᴇɴ ᴀ ʙᴀᴛʜʀᴏᴏᴍ ᴡᴀʟʟ is tiled, tiles of the same shape are fitted together to cover the flat surface without leaving any gaps, so that water cannot get through. We say that the tiles have been tessellated. In order for a shape to tessellate, the corners where they are brought together must equal 360°. For this reason, every triangle and every quadrilateral will make a regular tessellation, in which the pattern is based on just one flat shape. Regular hexagons also tessellate, as can be seen in honeycombs (p. 128). In an irregular tessellation, more than one shape is used. If you look at a soccer ball, for example, you will see that the patches are a combination of regular pentagons and regular hexagons, put together in a particular way so that they make a three-dimensional tessellation. Designs formed with tessellating shapes have featured in art and architecture for centuries, and the study of possible tessellations is used by scientists in the analysis of crystal structures (p. 153).

■ DISCOVERY ■
The mathematical artist

M.C. Escher (1898–1972) was a Dutch graphic artist best known for the strange optical effects of his work, which is sometimes called "mathematical art." In this woodcut, *Circle Limit IV*, the figures have no outlines. The contours of angels and devils define each other. To our usual way of seeing things, it looks as though the angels and devils are possibly tessellated over a sphere. However, the figures are simply decreasing in size by proportion, repeated forever, never leaving the circle.

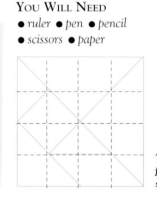

Circle Limit IV

EXPERIMENT
Tangrams

Tangrams are popular puzzles that were created in ancient China, and reflect the Chinese fascination throughout history with mathematical puzzles. The first record of a magic square (p. 26), for example, appeared at the start of the Han dynasty, around 200 ʙ.ᴄ. A tangram consists of a number of pieces in the shapes of triangles and quadrilaterals, which can all be cut from a large square. The puzzle is then to form the tangram pieces into various shapes representing stylized animals, people, and objects.

Yᴏᴜ Wɪʟʟ Nᴇᴇᴅ
● *ruler* ● *pen* ● *pencil*
● *scissors* ● *paper*

Template for tangram shapes

1 Dʀᴀᴡ ᴀ sǫᴜᴀʀᴇ on the paper and cut it out. Pencil in a grid of 16 squares to help you plot the shapes. Following the template shown above right, draw the tangram shapes on the paper square.

2 Cᴜᴛ ᴏᴜᴛ ᴀʟʟ sᴇᴠᴇɴ sʜᴀᴘᴇs with the scissors. You may like to leave the penciled grid lines on the shapes, so that you can easily tell which is the right side of the paper.

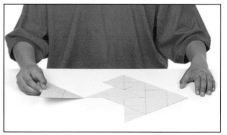

3 Mᴀᴋᴇ ᴀ sʜᴀᴘᴇ using all seven pieces. Shown above is a fish. See if you can make a cat or a face. On p. 187 there are examples of these shapes, and of other traditional tangram shapes.

Decorating with geometry

Simple shapes cut out of sponge can be used to make a wonderful tessellated pattern. The design needs to be planned first, taking into account the rules about what shapes will tessellate. The pictures on the right show two possible designs. Look around for inspiration from wallpaper, tiles, or art.

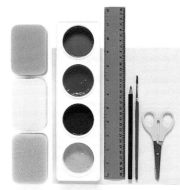

You Will Need
- *3 sponges*
- *poster paints*
- *ruler*
- *pencil*
- *paintbrush*
- *scissors*
- *paper*

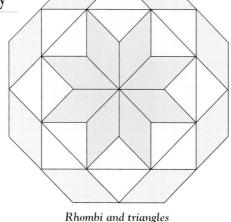

Rhombi and triangles
A rhombus (here, pink and green) has sides of equal length and two pairs of equal angles. A right-angled triangle can be seen as half a square. Put together in this way, the shapes form a regular octagon.

Octagons and squares
The octagons have equal angles and sides of equal length. The squares and octagons have sides of equal length.

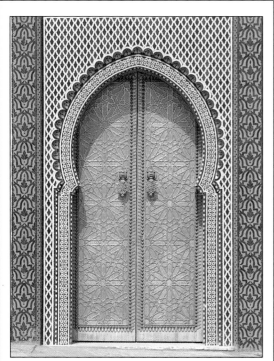

Tessellation in tiles
Islamic artists are skilled in the use of tessellations. In the Alhambra Palace in Spain, built for Muslim sultans, all 17 possible tessellating patterns have been found on the tiles. This mosque door in Fez, Morocco, has a complex pattern on the panels. There is also an underlying pattern in the straight-sided shapes linking the stars. The basic form is a right-angled triangle. Two of these form a small rectangle, and eight triangles make up a large rectangle. Each panel has three large rectangles. The tiles around the door contain tessellating patterns of curved shapes.

1 With a ruler and pencil, draw the shapes on the sponges. The lengths of the sides must be the same for both shapes. The picture shows a rhombus and a triangle. You could start by copying the examples above.

2 Cut out your chosen shapes with the scissors. Be careful to cut straight down and not at an angle, so that the shapes are not distorted on the base of the sponge.

3 Use a different sponge for each color. Dip the sponge shape into the paint and place it firmly on the paper. Pull the sponge off the paper with a clean action.

Make sure that the edges of the printed shapes touch without overlapping

CIRCLES AND WAVY LINES

FROM RIPPLES IN WATER TO THE MOON'S HALO, circles occur naturally throughout the Universe. Since the time of the Babylonians, people have been fascinated by the mathematics of this perfect form. Mathematicians have also studied other curves, such as the elliptical paths of planets and the parabolic paths of objects thrown into the air.

Over the last 4,000 years, many different types of curve have been discovered. Each has distinct properties that make it suitable for solving particular problems. Circles, for example, are used to make wheels, and curves known as "parabolas" are the basis for the shape of satellite dishes.

A perfect shape

Most of us are familiar with the circle. Like all other mathematical curves, it has a precise definition: the circle is a curve; both ends of this curve meet; and every part of its circumference (edge) lies at an equal distance from the central point.

In nature, circles are often formed when a flat surface is affected by a force acting equally in all directions. For example, when a pebble falls into water, it sends out a wave that spreads equally in all directions as a series of circular ripples.

People mimic this natural process when they build circular structures. For example, wagon wheels are made with spokes that pull all parts of the wheel's rim equally toward the center.

The ancient Greeks were fascinated by the circle. They believed it was a perfect form that could explain many mysteries of the Universe, such as how planets and stars moved through the heavens. They thought that the Earth was a perfect sphere. One scholar, Anaximander (611–545 B.C.), attempted to work out the longitude (distance from the equator) of his home. Four centuries later, another scholar, Eratosthenes (p. 135), managed

Conic sections
The ancient Greek scholar Apollonius, in his book Conics *(dated 225 B.C.), revealed that a circle-based cone can be cut to form curved shapes such as a circle, a parabola (above), and an ellipse (below).*

to make a fairly accurate estimate of the Earth's circumference. Astronomers in ancient Greece, such as Hipparchus (p. 126), were convinced that the Moon, stars, and planets moved around the Earth in circular paths. They devised elaborate models consisting of circles within circles, which could explain the observed motions of these bodies in the night sky.

Belief in the perfection of the circle survived for many centuries. In the Renaissance, for example, artists were often judged by their ability to draw a circle freehand.

Squaring the circle

Many ancient Greek scholars tried in vain to find important facts about the circle using geometry. In particular, they attempted to "square the circle" — to draw a set of shapes relating a circle's area to its diameter. The Babylonians also tried unsuccessfully to achieve this, but we now know that the area of a circle can never be found in this way. This is because the formula depends on the number π (p. 53). This is a special type of number, called a transcendental

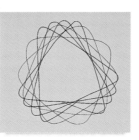

Tracing a curved path
This pattern is formed by paths within circles. To make it, a pen point is put through a round jar lid, then the lid is rolled inside a circle (p. 139).

number, that can never be written down exactly, nor can it be found by solving a mathematical equation or a geometric puzzle.

Ellipses

Most scientists accepted the ancient Greek idea that heavenly bodies move in circular paths, even when they realized that planets move around the Sun, not the Earth. However, in 1609 the German astronomer Johannes Kepler (1571–1630) made an interesting discovery. He found that all planets, including the Earth, had paths that were not circular but elliptical (in the shape of an ellipse). An ellipse is a circle stretched in one direction; it looks like the cross-section of an American football.

Part of a curve

A continuous piece of a circle, ellipse, or other curve is called an arc. Circular arcs have been used to build strong structures since

Curved patterns of sound
When radio waves are broadcast, a sound signal is combined with a "carrier" signal. In AM (amplitude modification), the sound signal varies the strength of the carrier waves. In FM (frequency modulation), it varies the waves' frequency. In this picture, an AM signal is shown on a spectrum analyzer as sharp curves of different sizes.

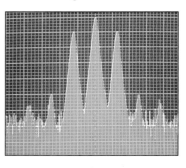

the time of the Romans. They used arcs, or arches, to support bridges and aqueducts (bridges for carrying water). The Romans realized that arcs could span longer distances and carry greater loads than lintels (straight, horizontal beams). Today, many bridges that span wide rivers or deep gorges do not use arches. Instead, they consist of flexible roadways that hang freely from strong cables. Such structures are called suspension bridges. The Humber Bridge near Hull, Britain, is one of the longest suspension bridges ever built: it is more than 4,626 ft (1,410 m) long. The earliest metal suspension bridges were built about 300 years ago, but many people, such as native Americans, have been building suspension bridges from other materials for many centuries.

All suspension bridges make use of a type of curve known as a "catenary" curve. If a heavy chain or wire is held horizontally but not pulled taut, it will sag a little to form a catenary curve (in fact, *catena* is the Latin word for "chain"). Overhead power cables naturally hang in this shape; so do the main cables that span a suspension bridge.

Curved paths

The parabola (p. 140) is the arc through which a ball moves when it is thrown into the air. When a ball is thrown, it gains both horizontal and vertical motion. As it travels upward, its vertical motion slows down due to the force of gravity. Eventually, the ball stops moving upward and begins to

A decoration made from a spiral
This decoration is made by taking a flat Archimedean spiral (p. 146) and stretching its ends apart.

plummet back down to Earth. The force of gravity makes the ball speed up as it falls. Parabolas are used to describe motion and also to help detect light, radio waves, and other forms of radiation. The insides of car headlamps, for example, often contain reflectors that are parabolic in cross-section. If light hits the reflector at any angle, it will bounce off to form a narrow beam straight ahead of the reflector. Similarly, a satellite dish is parabolic. Any radio signals that hit it are reflected as a beam, which is directed at a small detector in front of the dish.

Cycloids and cardioids (p. 144) are two interesting types of curve made by rolling circles. If you trace the path of one point on the edge of a circle that is rolling in a straight line, you produce a bump called a cycloid. If you trace the path of one point on the edge of a circle that is rolling around another circle of the same size, you make a heart-shaped curve called a cardioid. Many machines, such as presses and pumps, make repeating movements that are cardioid or

Collecting satellite signals
A satellite dish is a paraboloid, or a three-dimensional shape based on a parabola. Radio waves that hit this shape are reflected to the receiver (shown here to the left of the dish).

cycloid in shape. They contain parts attached to rolling wheels to produce this type of motion.

Spirals

A spiral is the shape you see from above as water disappears down a plughole. Some spirals, such as those made when cream is poured into coffee, are flat. Other sorts, such as spiral staircases, are three-dimensional. Mathematicians call a three-dimensional spiral a "helix."

A screw is a helix around a small metal cylinder. Screws were invented by Archytas of Tarentum (born 400 B.C.), a Greek philosopher. They are necessary in building work because they can hold two separate pieces of wood or metal together more securely than nails. The rotor blades of helicopters and propeller blades on airplanes cut through air in much the same way that metal screws cut through wood. This action propels the aircraft forward, and for this reason propellers are often referred to as "airscrews."

Springs are spirals that can be squashed or extended to store energy. This energy is released when the spring returns to its original size. Springs provide power in many instruments, such as mechanical clocks and watches.

Mathematicians try to use their understanding of spirals and helices to predict the effects of such shapes in the natural world, even those that occur in awesome, powerful forms such as a tornado (p. 148).

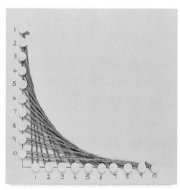

Curves from straight lines
Straight lines drawn on a graph can form curves like this parabola (p. 143). The lines, which run between the axes, are tangents to the curve. This pattern is made from string.

Discovering pi

PI IS THE RATIO of the circumference of a circle to its diameter, and is approximately equal to 3.14159. Pi is an irrational number (p. 38); this means that it cannot be expressed as the ratio of two whole numbers. However, the ratio 22:7 (or the fraction ²²⁄₇) is often used as an approximation for it, because the decimal representation goes on for ever. In about 3,000 B.C. the Babylonians gave pi as 3; later, the Greek mathematician Archimedes (p. 18) defined it more precisely as between 3⅐ and 3¹⁰⁄₇₁. A more accurate form has been sought, over the centuries, by people who include the Chinese, the Indians, and finally the Europeans. With the advent of computers, pi has now been calculated to many million decimal places. The symbol π for pi was first used in 1706 and was made popular by Leonhard Euler (p. 164) in 1737. It now occurs in all forms of mathematics, not just in geometry.

▓ Full Moon

Pi has been used by astronomers for centuries to calculate distances and orbits accurately. The Moon makes a full orbit of the Earth in about 29 days. When it is directly between the Earth and the Sun it is called a new moon. When it is exactly opposite the Sun, the sunlight falls on its nearest face to the Earth and we see a full moon. The perfect circle of the full moon is seen here in the night sky over the Gran Teton Mountains, Wyoming, United States.

▓ Puzzle

The circle of Sim is a deceptively easy game for two people. A circle is drawn, and six equally spaced points are marked on the circumference. Each player has a different colored pen and takes turns to draw a line joining any two points on the circle. Players must, however, avoid forming a triangle with their color, or they will lose. In the picture below, red has lost. The dotted lines show the only possible moves, each of which makes a triangle.

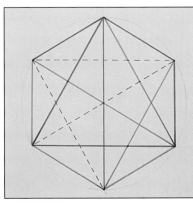

EXPERIMENT
Showing that pi is a constant

By working out the ratio between the circumference and diameter of several round objects, you can show very simply that π is always the same no matter what size of circle you are measuring.

YOU WILL NEED
● *calculator* ● *pen* ● *notepad*
● *tape measure* ● *circular objects,*
such as a saucer, cake tin, mug

If a tape measure is too unwieldy, use a piece of string

1 ON YOUR NOTEPAD write three column heads: circumference, diameter, and π. Choose a round object and measure its circumference. Write this down as accurately as you can under the appropriate heading. Measure the diameter and note this too.

2 DIVIDE the circumference by the diameter, and note the result under the heading π. Continue with other objects and compare your findings.

EXPERIMENT
Count Buffon's estimation of pi

Georges Louis Leclerc, Comte de Buffon (1707–88) was a naturalist and mathematician. He showed that if a needle is dropped from a random height onto a piece of paper covered with parallel lines, the length of the needle being equal to the distance between the lines, then the probability of the needle falling across a line is equal to ²⁄π. Copy his test and see if this formula works for you. (Check your calculation on p. 187.)

YOU WILL NEED
- *strip of paper 4 in (10 cm) wide*
- *pen* ● *notepad*
- *match* ● *ruler*

1 MEASURE THE MATCHSTICK, and make a note of the length. Lay the strip of paper vertically in front of you. Working from the top of the paper, draw lines across the strip, the same distance apart as the length of the matchstick.

Show each hit with a check mark and each miss with an ×

2 DROP THE MATCH on the paper. Ask a friend to note whether or not it falls across a line. Repeat at least 30 times. Count the numbers of check marks and ×'s. Calculate the practical probability by dividing the number of checks by the total number of attempts. Work out the theoretical probability as ²⁄π. Are the two numbers similar?

The circumference of the Earth

Eratosthenes (c. 240 B.C.) found the Earth's circumference by measuring the angles made by the Sun at noon in Alexandria, Egypt, and down a well in Syene, a known distance away on the same line of longitude. He measured the angle in the well to check that the Sun was vertically overhead (at an angle of 0°). The angle of the shadow at Alexandria at exactly the same time was 7°. He calculated that Alexandria must be about ¹⁄₅₀ of the circumference of the Earth from Syene (360° divided by 7°). The exact length of the units he used is not known, but one estimate of his calculation for Earth's circumference was 24,856 miles (40,000 km). Modern calculations put it at 24,870 miles (40,024 km). To find the Earth's diameter, divide this figure by π.

Finding the age of a tree

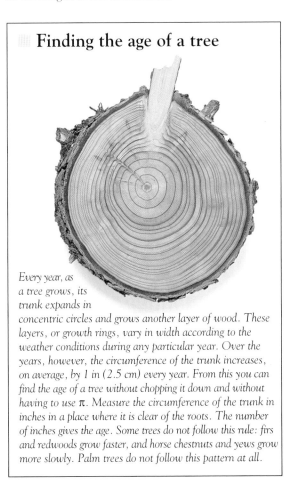

Every year, as a tree grows, its trunk expands in concentric circles and grows another layer of wood. These layers, or growth rings, vary in width according to the weather conditions during any particular year. Over the years, however, the circumference of the trunk increases, on average, by 1 in (2.5 cm) every year. From this you can find the age of a tree without chopping it down and without having to use π. Measure the circumference of the trunk in inches in a place where it is clear of the roots. The number of inches gives the age. Some trees do not follow this rule: firs and redwoods grow faster, and horse chestnuts and yews grow more slowly. Palm trees do not follow this pattern at all.

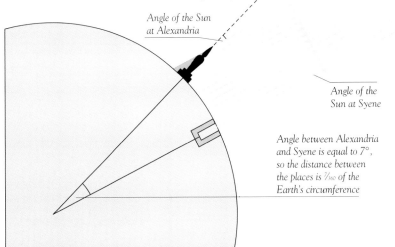

Angle of the Sun at Alexandria

Angle of the Sun at Syene

Angle between Alexandria and Syene is equal to 7°, so the distance between the places is ⁷⁄₃₆₀ of the Earth's circumference

Introducing the circle

A CIRCLE IS A SHAPE in which every point on its perimeter (circumference) is the same distance from its center. This distance is called the radius (r) of the circle. The line running through the center from circumference to circumference is called the diameter. The diameter equals twice the radius. Any straight line across the circle that does not pass through the center is a chord, and a part between two points on the circumference is called an arc. In mathematics, the circle is one of a group of curves that are known as conic sections (p. 140). Circles occur everywhere in nature: in the shape of the human eye, ripples on water, patterns on a butterfly's wing, and cross-sections of trees. A rainbow is a collection of circles, each one a different color. When we see the rainbow from the ground we see only the arc of each circle but, viewed from the air, the whole circle is visible.

Natural circles

The point at which a pebble hits the surface of a flat pond becomes the center of a number of circular ripples that spread outward as tiny waves. Eventually, the ripples die down as the energy supplied by the pebble is used up. The theory of waves and the mathematical formulas for them are crucial to physics, in fields such as fluid mechanics and magnetism.

EXPERIMENT
Find the center of a circle

This experiment shows an easy way to find the center of a circle without using any special instruments. You do not even need a pair of compasses to draw the circle — any circular object will do. By making marks on the circumference of the circle using a right-angled piece of paper, diameters can be drawn. Any two diameters of a circle cross at the circle's center. You can use this information to find the diameter and radius of a circle.

YOU WILL NEED
● *circular object,*
such as a plate
● *pen* ● *paper*
● *ruler*

2 LAY a smaller piece of paper over the drawn circle so that one corner touches the edge. Mark where the paper crosses the circumference, both on the circle and on the paper itself.

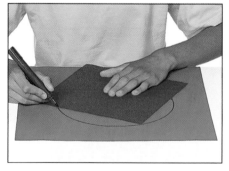

3 MOVE THE PAPER to another part of the circle, and position it so that the same corner and the marks on the paper meet the circumference. Go over the marks on the paper to put two new points on the circle. With a ruler, join the pairs of points across the middle of the circle so that the lines cross diagonally. The point at which they cross is the center of the circle.

1 PLACE THE circular object that you have chosen in the center of a piece of paper. Draw around the object carefully, holding it still with one hand and keeping the tip of your pen right up against the edge to make a smooth line.

EXPERIMENT
Finding the area of a circle

The area of a circle is calculated with the formula πr^2. It is also possible to work out the area roughly, without using π. In this experiment a circle is cut up and rearranged to make a rough rectangle. To find the area of this rectangle, multiply the breadth by the length. The breadth is equal to the radius of the circle and the length is roughly equal to πr.

YOU WILL NEED
- ruler ● pen
- pair of compasses
- scissors
- notepad ● paper

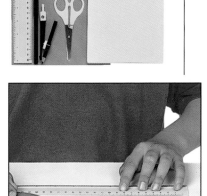

1 USING THE pair of compasses, draw a circle on the sheet of paper. Cut the circle out carefully. (You could work out the area of this circle now to see how close you get at Step 4.)

2 FOLD THE CIRCLE exactly in half, and make the fold sharper with your fingernail. Fold the paper in half again, and continue folding it into smaller and smaller sections as far as it will go.

3 OPEN the circle again, and flatten the folds. Cut the circle into pieces, following the fold lines accurately. You will eventually have a pile of wedge-shaped paper pieces.

4 LAY THE PIECES edge to edge as shown, to make a rectangle. Measure and write down the length and the breadth. Multiply these two numbers together to find the area of the rectangle.

EXPERIMENT
Patterns with circles and arcs

Mathematical principles can be used to make attractive objects. With a pair of compasses you can make patterns using arcs of a circle. An arc is part of the circumference of a circle. The arcs in this experiment all pass through the center of the original circle because the span of the pair of compasses is the same. You can try other patterns by setting the compasses to half the original span.

YOU WILL NEED
- pair of compasses
- paper

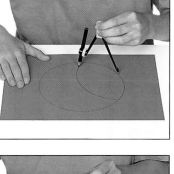

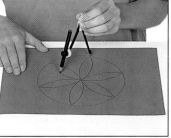

1 OPEN THE compasses and draw a reasonably large circle, with a radius of approximately 4 in (10 cm), at the center of the paper.

2 KEEPING the pair of compasses open at the same width, put the point anywhere on the circumference of the circle. Draw an arc that begins and ends on the circumference.

3 PUT THE point of the compasses at one end of the arc and draw a second arc. At the end of this arc, draw another one. Continue drawing arcs inside the circle until you return to the point where you started.

Straight-edged pattern with 12 points

Using the patterns
You can decorate key rings (above) or jewelry with circle patterns. For straight-edged designs, use a pair of compasses as in steps 2 and 3, but mark points instead of arcs. Rule lines across the circle to join the points.

Investigating circles

THE CIRCLE, CONSIDERED to be a perfect shape, is the locus (path) of a point moving at a fixed distance around another point. A circle encloses the maximum area for a given perimeter (p. 98). The analysis of circles can bring many practical benefits. For example, early studies allowed engineers to design the best shape for gear teeth so that they mesh perfectly as one revolves around another. In optics, the reflection of light in circular mirrors is studied, and circles provide the key to understanding the paths taken by light passing through lenses. The hands of a clock or watch move in circles divided into 60 equal arcs, each representing one minute. A circular protractor (p. 8) is marked with 360 equal arcs, representing 360°. And in photography, the amount of light a lens lets through to the film increases or decreases with the diameter of the circular aperture (opening).

■ Fairground thrills

As the Ferris wheel at a fair turns in the dark, the circles within its circumference can be clearly seen. When the wheel is moving at a constant speed, in every full turn it takes exactly the same time for each point within the circles to return to its starting place. Some points are farther from the center than others; these travel through longer distances, and so move faster. Points on the circumference travel fastest of all — a principle used to thrill riders at the fairground.

EXPERIMENT

Spots before your eyes

This experiment shows how order can be created, in the form of circles, from random dots. This is one example of mathematics with no instant practical use. However, it is now used by plastic surgeons to pinpoint the place for a new jaw hinge when a face is reconstructed. Random points are drawn on a picture of the jaw, then moved in the way that the jaw might operate. Circles appear, as in this experiment, and their center shows where the hinge should be.

YOU WILL NEED
● *acetate film*
● *pen* ● *push pin*
● *paper*

1 DRAW a large group of dots, in a completely random manner, on a sheet of paper. Make the dots fairly small — about the width of your pen's tip. Spread them over a wide area.

2 POSITION the sheet of acetate over the paper. (It may be helpful to stick both sheets down with masking tape.) Copy all the dots exactly onto the sheet.

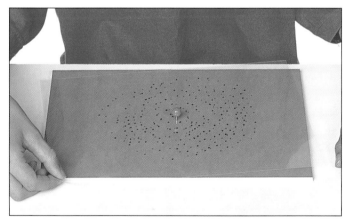

3 REMOVE ANY TAPE from the acetate sheet, then stick a pin through both the acetate and the paper sheets. Hold the paper steady with one finger, and turn the acetate sheet very slightly with the other hand. Watch the circles appear. Try this again, with the pin positioned in different places on the sheets, and you will see circles appearing around the pin.

EXPERIMENT
Drawing circles within circles

It is possible to follow the paths of points within a circle as well as those on the circumference. The simple mechanism in this experiment gives a visual demonstration of how points inside a small circle travel inside a larger circle. The study of these shapes has interested many mathematicians, partly because of the shapes' purity and partly for their practical applications in relation to gears and planetary movements.

YOU WILL NEED
● *corrugated cardboard* ● *ruler*
● *pens* ● *pencil* ● *pair of compasses* ● *craft knife* ● *scissors* ● *paper* ● *14 in square (35 cm square) foamcore* ● *cutting mat* ● *adhesive tape* ● *small bradawl or screwdriver* ● *jar lids* ● *glue*

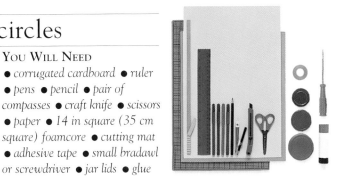

Adult help is advised for this experiment

1 IN THE CENTER of the foamcore, draw a circle with a diameter of at least 10 in (25 cm). Place the foamcore on the cutting mat, then ask an adult to cut out the circle for you with a craft knife.

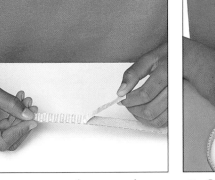

2 CUT LONG STRIPS of corrugated cardboard about ½ in (12 mm) wide. Carefully peel the layer of covering paper off one side of each strip, to expose the ridges inside the cardboard.

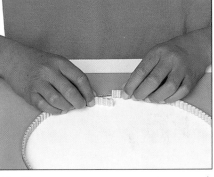

3 GLUE THE CORRUGATED strips around the edge of the circle. To add a new strip, tear a bit of the covering paper off both sides of one end and overlap two or three ridges on the strip underneath.

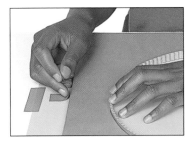

4 TO PREPARE the frame for use, put a sheet of paper under the foamcore. Secure the paper to the table with adhesive tape, and attach the foamcore firmly to the paper and the table.

5 ASK AN ADULT to make two or three holes in each jar lid, at different distances from the edge. Pierce the holes from the outside to leave the bases smooth. Flatten any rough edges inside the lids.

6 CUT some more strips of corrugated cardboard, and glue them around the outer edge of each jar lid. Cut them to length, and overlap the ends of the strips carefully as described above.

Making designs in a circle
Choose a jar lid, and lay it upside down inside the circular frame. Position the lid against the edge of the circle so that the two sets of cardboard teeth interlock. Choose a pen, and hold it upright with the tip stuck through one of the holes in the lid. Then roll the lid around the circle, guiding it gently with your free hand so that the teeth interlock properly without wobbling. The pen should trace a path around the circle. Keep the pattern going for as long as you can. What do you notice? What other sorts of patterns can you make when you use different holes or different lids?

Introducing curves

A CURVE IS A LINE that continuously bends but has no angles (p. 116). There are many different sorts of curve. Some, such as parabolas, are open; the line never returns to its starting point. Some do join up with themselves, such as ellipses; these are called closed curves. Some curves are twisted, like the helix (p. 148). A heavy chain left to hang between two points forms a natural curve known as a catenary curve (p. 133). The path made by a point on a wheel as it rolls along a flat surface is a cycloid (p. 144). The shape of an aircraft wing (p. 107), bridge structures, and the paths of rockets are all determined by the mathematics of special curves. Some groups of statistics can be analyzed (p. 82) to see if patterns of data repeat themselves to form curves. Many natural phenomena follow curves — the flow of a liquid and the path of a thrown object, for example. Algebraic functions plotted on graphs often appear as curves (p. 75).

Juggling

Watch someone throwing a ball or juggling, or try it yourself. As each ball is thrown into the air, the force of gravity acts on it to conteract the original upward force. Because of this, the ball will follow a parabolic curve.

EXPERIMENT
Cutting through cones

This experiment shows how to produce a number of curves by cutting through a circular-based cone at different angles to the base. These shapes are known as conic sections. A circle is produced by cutting through the cone parallel to its base. A parabola is produced by cutting parallel to a sloping side, and an ellipse is made by cutting through the cone at a slant. Apollonius, the Greek geometer, was the first to show this method in his work on conics in about 200 B.C.

YOU WILL NEED
● scissors ● modeling clay
● spool of thread ● 2 corks

1 CAREFULLY roll the modeling clay between your hands and on the tabletop to form a cone shape. Try to make the shape as accurate as possible. (You can use a circular object as a guide for forming the base.)

2 CUT A 10-IN (25-CM) length of thread. With the scissors, make small notches halfway down the sides of both corks. Tie the ends of the thread to the corks, lodging them securely in the notches so that they do not slip.

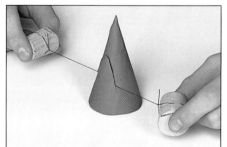

3 STAND THE CONE in front of you. Hold the thread taut, and pull it through the cone to cut smooth sections. In this picture, the side of the cone is cut vertically to form a hyperbola (an open curve with two symmetrical branches).

Parabola
This is formed by cutting a section through a circle-based cone parallel to a sloping side.

Ellipse
This is formed by cutting a section through a circle- or ellipse-based cone at a slant.

Receiving radio waves

Satellite dishes are used to collect radio waves from satellites in orbit around the Earth. Some, called radio telescopes, can also be used to pick up radio waves from distant planets and stars. The dishes are paraboloids, solid shapes that have a parabolic cross-section. Whenever a radio wave traveling parallel to the axis of the dish strikes the surface, the paraboloid shape causes it to be reflected directly back to the focus (p. 183). A receiver is positioned at the focus. The receiver picks up the waves, and is connected to a computer or to your television set.

(p. 183)

EXPERIMENT
Planetary orbits

Planets in the Solar System orbit the Sun in a movement that is elliptical rather than circular. The lengths of their orbits, and the speeds at which the planets travel, depend on their distance from the Sun. By carrying out this experiment, you can understand how planets that are close to the Sun — with relatively short orbital paths — travel at different speeds from planets that are farther away.

YOU WILL NEED
- *long and short rulers* ● *glue*
- ● *modeling clay*
  - ● *adhesive tape*
  - ● *scissors*

1 LAY THE two rulers on the table, with their ends aligned. Secure each ruler to the tabletop with adhesive tape. Stick equal-size balls of clay to the other ends of the rulers.

2 HOLD both rulers upright. Let them both go at the same instant, so that they fall forward (away from the tape at the bases). Watch to see which ruler falls to the table first. Do this a few times to confirm your result.

There should be a small space between the bases of the rulers

---

EXPERIMENT
Drawing an ellipse

It is difficult to draw an ellipse by hand, but this simple technique produces fairly accurate shapes. The pins represent the two foci of the ellipse (p. 183). In an ellipse the total distance from one focus to any point on the curve, and from that point to the other focus (the length of the string), is always constant.

YOU WILL NEED
- ● *adhesive tape* ● *pencil*
- ● *push pins* ● *string* ● *scissors*
- ● *colored paper* ● *foamcore*

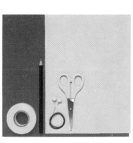

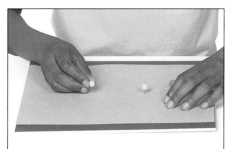

1 LAY THE COLORED paper on the foamcore, which will provide a firm base in which to stick the pins. Push two pins a small distance apart through the colored paper and into the foamcore.

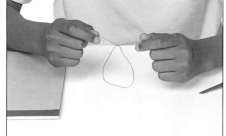

2 CUT A LENGTH of thread long enough to go round the two pins, plus a little extra. Lay it around the pins to form a loose loop, then lift it off and tie a double knot where the ends cross.

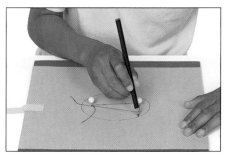

3 DROP THE LOOP of thread over the pins. Put the point of a pencil inside the loop. Move the pencil around the pins, keeping the thread taut so that the line is as smooth as possible.

Special curves

MANY TYPES OF CURVE, such as the parabola (p. 140) and the catenary curve (p. 133), occur in nature. Scientists studying these curves have discovered that they can offer highly effective mathematical solutions to given problems. For example, cardioid curves (p. 144) were studied by people designing teeth for gears. One surprising tool for exploring curves is soap. A film of soap stretched between shapes will form the minimum possible surface area to enclose a given volume, and shows the shortest distance between those points that it joins. Chemists use soap bubbles to find out how atoms are arranged in some molecules. Engineers, and town and road planners, also use the mathematics of soap film in many of their analyses.

Using simple technology

The stadium for the Munich Olympics (1972) was designed by Frei Otto, a German architect. He created the building from strong, lightweight structures that could be assembled and disassembled easily, and achieved this by imitating the curves in soap films.

Spectacular shapes in soap

Mathematicians are fascinated by soap films because they can help to solve complex problems of space very simply. Because of the tension within the liquid, a soap film will always form the smallest surface area between points or edges. A film stretched across a hoop, for example, will form a flat disk, and a film around a volume of air will form a sphere (a bubble). When shapes, or groups of points, are dipped into soap and lifted out again, the film reveals useful information.

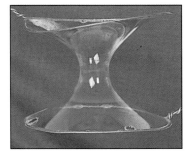

A catenoid
This shape (p. 182) is made with two separate wire circles of equal size. Hold the circles together, dip them into soap mixture (right), then slowly move them apart to form a funnel.

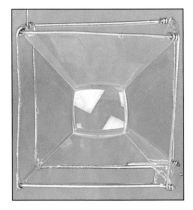

The curved cube
If a cubic frame (right) is dipped into soap mixture, the film will stretch between the edges to form a shape like this. The straight-sided, double funnel, with the curved, cubelike shape at the center, joins all the edges of the cube with the minimum amount of soap film.

EXPERIMENT
Making a frame for soap film

In this experiment you make a cubic frame from wire, and dip it into soap mixture to see the minimum surface area that will join the edges. You could also make frames for other shapes such as hoops (left), a tetrahedron (p. 152), or a shape of your choosing. A good soap mixture can be made from 11 parts water, 5 parts liquid detergent, and 1 part glycerin (which strengthens the film and helps to prevent it from bursting). Use the same unit of measurement for all three liquids. The ratio 11:5:1 (p. 56) applies no matter how much soap mixture you make.

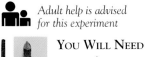 *Adult help is advised for this experiment*

YOU WILL NEED

- *wire cutters*
- *wire*
- *paper*
- *pen*

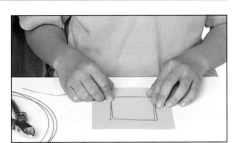

1 DRAW A 4-IN (10-CM) square on the paper, to form a template for the wire shape. Bend the wire around the shape, and secure it at one corner by twisting one end of wire around the other. Make two wire squares in this way.

2 MAKE FOUR UPRIGHT sections from equal lengths of wire slightly longer than the sides of the template. Use them to join the squares together at each corner to form a cube. Ask an adult to cut off any loose ends. Make a wire holder.

YOU WILL NEED
- ruler • pencil • scissors
- string or thick thread
- thumbtacks • foamcore

EXPERIMENT
Making curves with straight lines

It is possible to make curves by drawing straight lines, on a graph with two axes (p. 75). Lines drawn between the axes cross each other, and the outer surface of this pattern forms a curve. The straight lines are actually tangents to the curve (p. 185). This experiment uses a board marked in the same way as a graph, with two axes at right angles to each other and with numbered points. Thumbtacks are stuck into these points, and thread or string is wound around them to make the straight lines. There should be the same number of points on each axis; the numbering starts at the top of the vertical axis, and at the left of the horizontal axis. (No number is written at the corner of the graph.) Axes at 90° create a shallow parabola, while axes drawn at an acute angle form a steeper one. You may also like to combine different patterns.

1 DRAW THE HORIZONTAL and vertical axes on the foamcore, making both the same length. Mark the axes with a scale of your own, say intervals of 1 in (2 cm), and number the marks along both axes.

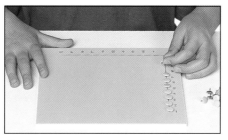

2 INSERT THUMBTACKS along both axes, at every mark on the scale. If you extend the scale to the edge of the board in both directions, use the same number of pins on each axis.

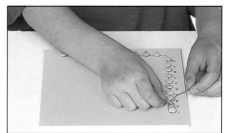

3 UNROLL A LONG PIECE of thread or string. (Do not cut it.) Hook it around the thumbtack at the top of the y axis, or vertical axis (number 1), and secure the end with a small knot.

DEMONSTRATION
Painting with a pendulum

Modern art using curves
Cut the base off a liquid detergent bottle, make four holes near the cut edge, and tie string through them. Make sure that the cap is on. Fill the bottle with thin liquid paint. Hang it from a hook well away from furniture or walls, with a large sheet of paper under it. Hold the bottle at about 45° and remove the lid. Gently let it fall, and stand back. The pendulum will swing in natural elliptical curves (p. 140).

The pendulum makes a different curve on every swing

4 LOOP THE STRING around the first thumbtack (1) on the x axis, then back to the second thumbtack (2) on the y axis. Pull the string straight, but take care not to loosen any of the thumbtacks.

5 CONTINUE TO WRAP the thread around thumbtacks in sequence until you reach the last number on each axis. Steady the tacks with your finger so that they do not come loose.

Uses for curves

DURING THE GOLDEN AGE of Greek mathematics, about 300 to 200 B.C., three mathematicians were particularly important: Euclid (p. 114), Archimedes (p. 18), and Apollonius (*c.* 260–190 B.C.). Apollonius wrote a work on conics (p. 140). There was not much practical benefit for his work until centuries later. Galileo (p. 28) used the study of cones in his work on the paths of projectiles, and the astronomer Johannes Kepler (1571–1630) showed that the orbit of a planet is in the shape of an ellipse (p. 141). The parabola is used in satellite dishes (p. 141), and car headlights, in which the light bulb is placed at the focus of a parabolic mirror so that all the light is reflected in straight lines onto the road. A brachistochrone is a curve that defines how to travel from a high point to a lower point, not directly underneath, in the shortest time. It has been shown that a brachistochrone is always a cycloid.

Le Pont de Normandie

This suspension bridge spans the Seine estuary in Normandy, France. The decks of the bridge are designed as special curves so that the wind flows across them as easily as over the body of an aircraft. The bridge has a central span of 2,800 ft (856 m). The span is so long that it has been built with a slight upward curve from end to end, to accommodate the effects of the curvature of the Earth. The huge cables suspend the bridge in the air.

EXPERIMENT
Creating new shapes with circles

This experiment demonstrates how to form a cycloid and a cardioid. These two shapes are the loci (paths) of a point on a circle as it travels along a straight line (cycloid), or around another circle of the same size (cardioid). Both of these shapes have practical applications. For example, a cycloid is the strongest shape for the arch of a bridge, and cardioids have been used in the search for the best shape for gear teeth.

YOU WILL NEED
● *ruler* ● *pen* ● *putty adhesive* ● *scissors* ● *2 rolls of colored adhesive tape* ● *2 circular objects of the same size, such as jar lids* ● *paper*

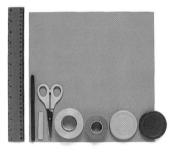

Drawing a cycloid
Secure a piece of paper to the table with putty adhesive. Tape a ruler to the paper, parallel with one long edge. Cut a diamond shape from colored adhesive tape, and stick it to one jar lid with the point touching the edge. Lay the lid against one end of the ruler, with the diamond pointing down. Mark this point with a pen. Then roll the lid along the ruler, marking the diamond's position at frequent intervals.

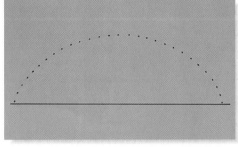

Finished cycloid on straight line

Drawing a cardioid
On another sheet of paper, fix the unmarked jar lid to the center with putty adhesive. Draw around this circle with a pen. Lay the marked lid against it with the diamond pointing at it. Make a dot by the point of the diamond. Roll the marked lid around the first lid, marking the position of the diamond at frequent intervals. You should end at the point where you started.

The curved line of dots should start and end at the same point on the circle

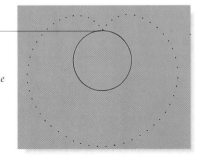

Finished cardioid around central circle

EXPERIMENT
Making a brachistochrone

By rolling two table tennis balls on this model, you can see that the curve of a brachistochrone — a cycloid-shaped path — is the fastest route from one point down a slope to a lower one, even though the curve is longer than a straight path to the same point. Why do you think this is true?

YOU WILL NEED
● *large box* ● *corrugated cardboard* ● *ruler* ● *glue* ● *pen* ● *double-sided adhesive tape* ● *scissors* ● *adhesive tape* ● *paper* ● *round dish* ● *2 table tennis balls* ● *protractor*

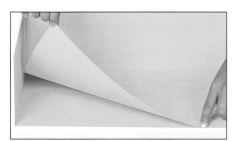

1 FIND a dish with a circumference roughly equal to the box length. Align the dish with a long edge of the paper and draw a cycloid (see below left).

2 WITH the protractor at the left-hand end of the cycloid, mark an angle of 10°. Draw a line at that angle across the paper.

3 CUT ONE LONG SIDE off the box. Then trim the paper with the cycloid to the correct size, and glue it to the back of the box, with the curve toward the base and the straight side aligned with the top of the box.

4 CUT A 2-IN (5-CM) wide corrugated cardboard strip to the length of the cycloid. Score the paper across the back of the strip so that it bends, then tape it to the sides of the box to follow the cycloid line across the back of the box.

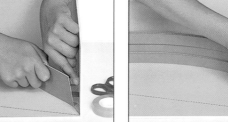

5 CUT TWO MORE cardboard strips to the same length and ¾ in (2 cm) wide. Score the backs. Align them with the edges of the larger strip, and secure them with double-sided tape. These strips form the sides of a groove for the ball.

6 MEASURE the length of the straight line inside the box. Cut a cardboard strip to this length, and 2 in (5 cm) wide. Cut two more strips ¾ in (2 cm) wide. Tape these to the edges of the wider strip.

Using the brachistochrone
Hold the two table tennis balls at the tops of the tracks. Let them both go at the same moment, and watch them roll along the tracks. Which travels faster? You might like to show this model to a friend. Before you release the balls, ask your friend to guess which one will reach the stop first.

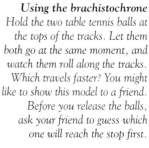

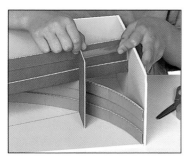

7 TAPE the straight strip inside the box so that it follows the ruled line. Cut a 4×2 in (10×5 cm) cardboard strip. Tape it across both tracks at the point where they meet, to stop the balls.

The nature of spirals

A SPIRAL IS A SPECIAL TYPE of curve that occurs in two dimensions, on one plane (p. 183). There are several types of spiral, which are given different names according to their shapes and the algebraic equations that define them (p. 75). Archimedes (p. 18) studied spirals, and one particular spiral is named after him. Some shells, such as that of the nautilus (p. 59), have a spiral shape. Spirals are found in many places — for example in the centers of whirlpools, and the vast movements of hurricanes as shown from above in satellite pictures. Some of the galaxies in the Universe are described as spiral, because their shapes have spiral arms. (Others are described as elliptical galaxies, or as irregular galaxies if they have no definite shape.)

Spiral galaxy

Our own Milky Way is a spiral galaxy — one of around 100,000 million galaxies in the universe. This picture shows a spiral galaxy known as M74, in the constellation of Pisces. It is approximately 30 million light-years away from us (p. 42) and 80,000 light-years across. Spiral galaxies differ in the shapes of their spiral arms and the sizes and shapes of their centers.

EXPERIMENT
Drawing an Archimedean spiral

One particular type of spiral is known as the Archimedean spiral. It is the path of a point that is either coming toward or going away from an origin at a constant speed, while also moving around it at a constant velocity. There are many ways to draw the spiral. This experiment shows why the Archimedean spiral is this particular shape.

YOU WILL NEED
● *ruler* ● *pen*
● *adhesive tape*
● *scissors* ● *spool of thread* ● *a large piece of paper and a small piece*

1 MEASURE a strip of paper the width of the spool and 10 in (25 cm) long. To find the width of the spool, measure just the smooth part, excluding the rim, so that the paper will wrap around the spool without snagging.

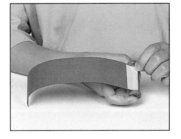

2 CUT OUT the paper strip and stick one end to the spool of thread with adhesive tape. The paper should be the same width all along, and the end should be parallel to the edges of the spool, otherwise it will snag on the rim.

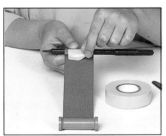

3 LAY THE spool of thread with the tape beneath it. Position the pen at the other end of the paper strip. Fold the paper around the pen. With adhesive tape, attach it about 1 in (2 cm) from the tip of the pen.

4 TURN THE strip around with the spool toward you. Roll the strip tightly around the spool, up to the pen.

Your friend should hold his hand clear of the spool so that the pen can pass underneath

5 ASK a friend to hold the spool steady in the center of the large sheet of paper. Hold the pen upright and pull it gently so that it draws as you unravel the coiled strip of paper. You will have to change hands carefully as you draw past your friend's arm.

Moving spiral

If you hang a paper Archimedean spiral over a lamp, an upward flow of warm air created by the heat of the light bulb will make the spiral turn. Rising currents of warm air in the Earth's atmosphere are called thermals, and birds and gliders use them to fly upward seemingly without effort.

YOU WILL NEED
- *an Archimedean spiral on poster board (below left)*
- *string* • *scissors* • *lamp*

Make sure the end of the spiral is clear of the lamp

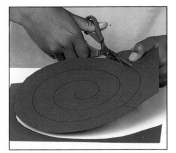

1 DRAW AN Archimedean spiral (see below left). Starting at the outer edge of the spiral, cut it out along the drawn line up to the end.

2 USING the scissors, make a hole through the center of the circle. Thread the string through it, and secure the string with a double knot.

3 SUSPEND the decoration above the lamp and then turn the light on. The spiral will start to move because of the warm air rising.

A Celtic spiral

People have been fascinated by spirals for centuries, and have depicted them in art. Among them were the Celts, who migrated across Europe during the Bronze Age. Shown here is the Celtic artists' method for drawing regular spirals without the aid of complicated equipment. You do not need compasses to make the outer circle — any round object 9 in (24 cm) in diameter will do.

YOU WILL NEED
- *T-shirt* • *ruler*
- *pair of compasses*
- *fabric paint*
- *poster board*

1 FIND a point on the T-shirt where the center of your chest would be if you were wearing it. Using compasses or a flat, round object, draw a circle with a radius of 4¾ in (12 cm).

2 LAY a ruler horizontally on the circle, level with the center. Draw a line, and mark ⅝-in (17-mm) intervals from one end of the line to the other to divide the line in 14 parts.

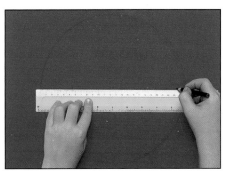

3 STARTING AT THE FOURTH MARK from the left, paint a spiral that goes above the line, passes down through the next point to the right, then up through the next to the left, and so on to the edge. In another color, start a spiral four marks from the right. Make it go under the line, up through the next point, and so on.

Three-dimensional spirals

A HELIX IS A THREE-DIMENSIONAL SPIRAL that looks a little like a corkscrew. Unlike the Archimedean spiral (p. 146), which is flat, the helix is described mathematically as a line going at a constant angle around a cylinder or a cone. Helices occur in both technology and nature, most notably in the double-helix structure of DNA, discovered in 1953. The waste metal, or swarf, that a drill leaves when it cuts through soft metal often takes the form of a helix. A "spiral" staircase is actually helical. An early example of a helical spiral is Archimedes' screw. Invented by Archimedes (p. 18), this giant screw was used to raise water from low-lying areas. This principle is also used in harvesting. Grain collected in a combine harvester is drawn up a tube, in the same way as water in Archimedes' screw. It falls out of the top and into a truck alongside the harvester.

Natural spirals

A tornado is the most violent kind of storm, and usually leaves total devastation in its path. It is an intense whirlwind shaped like a funnel, with very high winds spiraling toward its core. The funnel-shaped cloud is visible because of the dust and water droplets that are sucked into the center. The helical spiral usually spins counterclockwise in the Northern Hemisphere. Waterspouts (tornadoes over water) are also helical in shape, as are hurricanes, powerful storms that affect very large areas.

EXPERIMENT
Spring balance

A spring balance is a weighing instrument that compares weights. Springs change shape in a regular way. If you push some types, they compress. If you pull other springs, they stretch. In the experiment, a helical spring stretches a certain amount with one weight attached; watch what it does when the load is doubled.

YOU WILL NEED
- pail ● string ● tape measure ● spring
- pencil ● notepad ● 2 identical weights

1 FIX THE spring to a door frame or tie it to a secure place with a piece of string. Tie the pail to the spring with another piece of string. Measure the length of the spring.

2 PLACE A weight (such as the heavy padlock used here) in the pail. When the spring is still, measure how much the spring has extended or stretched.

3 PLACE A SECOND identical weight in the pail and measure the length of the spring again. What do you notice about the difference between the spring when it holds one weight and the spring as it holds twice the weight? (Answer on p. 187.)

EXPERIMENT
Making a double helix

Almost all organisms, including human beings, have a genetic "fingerprint" determined by their DNA. The DNA molecules hold information about an organism's inherited traits, and these molecules copy themselves by "unzipping" down the middle and regenerating their missing half. In this experiment you can make a simplified model of a DNA molecule, showing all its components. If you look closely, you can see that the bases make a further helix, in addition to the two on the outside.

YOU WILL NEED
- *inner tube of a paper towel roll*
- *pen* • *wire*
- *modeling clay in 2 colors* • *toothpicks*
- *adhesive tape*
- *scissors*

■ DISCOVERY ■
Nobel Prize winners

At Cambridge University, England, in 1953, chemists Francis Crick and James Watson made a breakthrough in genetics. They used information about the structure of proteins and X-ray pictures made by Rosalind Franklin (1920–58), as the basis for a chemical model they created that showed the DNA molecule as a double helix. Using the model, Crick and Watson determined how DNA could copy itself and carry inherited characteristics from one cell to another.

Each clay ball should be about as big as a cherry

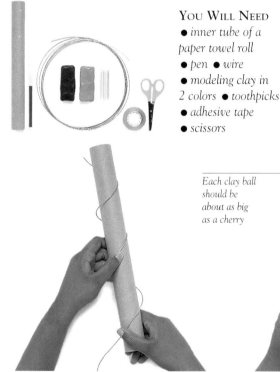

1 USING THE kitchen-roll tube as a template, curl the wire around it in a spiral. Follow the seam on the tube to keep the coils evenly spaced. Make another identical spiral.

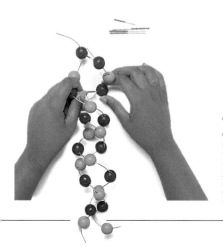

2 MOLD THE modeling clay into 12 balls in each color. These balls represent the sugars and phosphates on the outside of the molecule. Thread them onto each helix, alternating the colors as you go. Make sure that both helices start with the same color. Wrap strips of adhesive tape around the wire above and below the balls, to hold them in place.

3 CUT THE ENDS off the toothpicks, then color half of each with the pen. These two-colored sticks represent the pairs of chemical compounds (bases) that join the helices together. Line up the wires so that they intertwine, then use the sticks to link each pair of clay balls of the same color. The sticks should run horizontally across the center of the model.

Two chains of sugars and phosphates make the double helix

The bases are attached to the sugars, and make another helix within the model

On this model the colors on the bases alternate down the spiral, for visibility, but in reality the bases combine in regular but complex arrangements

The finished DNA spiral
This model is simplified. In reality each chain looks like a ribbon. There are four types of bases, joined by hydrogen bonds and forming a complex pattern.

3D AND SYMMETRY

OUR WORLD IS MADE UP of three-dimensional objects. Some, like the Earth, are curving, while others, such as crystals, are made of flat shapes that fit together. Over the centuries, mathematicians have found many ways to classify these shapes, and ways of looking at them, such as topology, which have given us a deeper understanding of space.

Every object we can see exists in three dimensions: it has a height, a width, and a depth. Mathematicians call three-dimensional objects "solids."

Some solids, such as our own bodies, have continuously curving surfaces. Others, such as cubes and pyramids, are made of polygons (flat, straight-sided

Great Pyramids
The ancient Egyptians had a very limited knowledge of theoretical geometry but managed to create the huge, regular, square-based Pyramids at Gizeh.

shapes) that fit together exactly. Mathematicians call a solid made from polygons a "polyhedron."

Perfect solids

The Pythagoreans of ancient Greece (p. 124) found that some polyhedra could be fitted inside a sphere. These solids had faces that were all regular shapes (p. 128), such as squares or equilateral triangles. The Pythagoreans called these solids "perfect solids." The simplest perfect solid they identified was the cube. Others were the regular tetrahedron, the regular octahedron, the regular

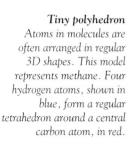

Tiny polyhedron
Atoms in molecules are often arranged in regular 3D shapes. This model represents methane. Four hydrogen atoms, shown in blue, form a regular tetrahedron around a central carbon atom, in red.

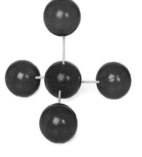

dodecahedron, and the regular icosahedron (p. 152). We now call these five solids "Platonic solids," after the ancient Greek philosopher Plato (429–348 B.C.) who tried to explain the physics of the Universe by studying them.

Solids in nature

Scientists have discovered that the atoms in many molecules are arranged in the shapes of Platonic solids. All substances consist of atoms, or groups of atoms bonded together to form molecules. A salt crystal, for example, consists of sodium and chlorine atoms linked in a cubic structure. The atoms in a methane molecule (below left) form a tetrahedron shape.

A recently discovered form of carbon, buckminsterfullerine, has a very interesting structure. Each molecule of the substance consists of 60 carbon atoms linked to form a ball. This structure, known as a "buckyball," looks like an icosahedron that has had its corners cut off.

Buckminsterfullerine was named after the British environmentalist and architect Richard Buckminster Fuller (1895–1983). He predicted that buildings based on icosahedrons would be extremely strong and lightweight. Scientists have since realized that buckyballs possess both these properties. They have also discovered that

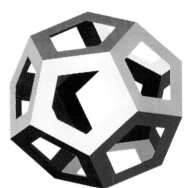

A perfect polyhedron
The dodecahedron is one of the five "Platonic solids" (p. 152). It has 12 faces, each one a regular pentagon.

buckminsterfullerine, in certain conditions, is a "superconductor," offering little resistance to electrical current. Engineers hope to use superconductors in compact, powerful computers.

Changing shapes

Mathematicians have always been interested in making new shapes by altering existing ones in certain ways. This process is called "transformation" (p. 156).

Objects seem transformed when we look at them from different angles. The rim of a mug, for example, looks circular from above but elliptical when we see it at an angle. For this reason, artists distort the actual shapes of objects in their pictures to give the impression of three dimensions on a flat surface.

Gerardus Mercator
Mercator devised the Mercator projection (p. 160). This form of map projection is a valuable aid to navigators, and has been widely used for several centuries.

The French mathematician Girard Desargues (1593–1662) was one of the first people to show geometrically how objects should be drawn to look three-dimensional. This aspect of art is called perspective. Our understanding of perspective has led to the development of mathematical projections.

N. I. Lobachevsky
This Russian mathematician (1792–1856) published the first work to take geometry beyond Euclid, and paved the way for work on topology and the fourth dimension.

Images of solids

A projection is a solid object represented on a two-dimensional surface. Maps and elevations (side-views) of buildings are examples of projections.

World maps represent the near-spherical Earth on flat sheets of paper. Most world maps use a "Mercator Projection" (p. 160). This form of projection has a few drawbacks. In particular, it makes countries near the poles, such as Canada and Australia, look far too big and countries near the equator, such as India, look too small. Map-makers have also developed other projections such as the Peter's Projection. This distorts the shapes of land masses a little but accurately shows their area.

Symmetry

Regular patterns can be classified using symmetry (p. 158). This occurs in natural as well as geometric forms. For example, a butterfly, seen from above, shows symmetry along a vertical line through its body. The pattern on one wing is a direct reflection of that on the other wing. This is known as "bilateral symmetry."

A shape that has "rotational symmetry" looks the same if it is turned through a particular angle. All regular polygons and regular polyhedra have rotational symmetry, as do snowflakes and the petals of some flowers.

Groups of tessellating objects (p. 130), such as bricks in a wall, and natural forms such as lumps of quartz crystal, have "translational symmetry." They look the same if they are moved through the distance between the centers of each tessellating shape.

Topology

In the 19th century, the French mathematician Jules Henri Poincaré (1854–1912) studied shapes in a completely new way. In his studies, surfaces could now be distorted, but not punctured or mended, without losing their essential form. So a triangle (a shape with no holes) has the same essential form as a square or a vase; each could be made by reshaping the other. Similarly, a mug with a handle (a shape with one hole) has the same form as a doughnut with a hole in the middle.

This discipline, called topology, is now applied in many fields outside mathematics — for example, to analyze complicated strings of DNA or to design circuits for silicon chips.

Distorted space

Topology can be used to relate Euclidean geometry (p. 114) to new geometric forms. In the last 150 years, mathematicians such

Modern technique
The Space Shuttle was designed on a computer, using a program that allowed the plan to be shown in three dimensions. This technique is now used in designing machines, vehicles, and buildings.

as Nicolai Lobachevsky (above), János Bolyai (1802–60), and Georg Riemann (1826–66) have developed systems of geometry in which the facts assumed in Euclidean geometry are changed. Their systems, which are called "Non-Euclidean geometries," are the mathematics of seemingly strange spaces where the simplest surfaces are curved rather than flat. In the 20th century, physicists have used these systems to understand the properties of space, in particular the distorted space that creates black holes.

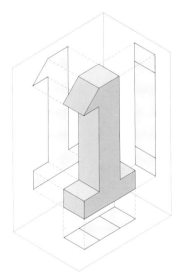

Projecting on to a flat surface
Projections are used in architecture and industrial design to show the faces of a 3D object on a 2D surface such as a piece of paper or a computer screen.

A compact shape
Despite its intricately curved shape, the French horn has just one outer surface and one inner surface. Air is displaced through valves to produce the notes.

Solid shapes

THE ANCIENT GREEKS had a particular interest in the study of solid shapes, and we use their word *polyhedron* to define a solid figure with four or more flat surfaces. The main features of a polyhedron are: faces, which are the surfaces; edges, the lines where the faces join; and vertices, the points where the edges join. The more faces there are on a polyhedron, the more it resembles a sphere. Knowledge of the mathematical properties of solid shapes is vital in the study of crystals and molecules. For instance, some minerals can be identified simply by the shapes of their crystals and the way they fit together. New polyhedra are still being discovered — these include complex structures that can be collapsed into single, flat surfaces.

▓ Platonic solids

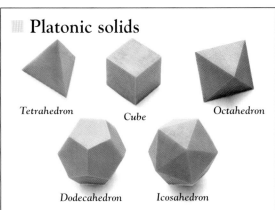

Tetrahedron　　　*Cube*　　　*Octahedron*

Dodecahedron　　　*Icosahedron*

These shapes are regular polyhedra. All the faces are one type of regular polygon (p. 128), such as an equilateral triangle or a square, and all the vertices are congruent (have identical angles). The tetrahedron has four triangles; the cube has six squares; the octahedron has eight triangles; the dodecahedron has 12 pentagons; and the icosahedron has 20 triangles. Euclid (p. 114), in his books *Elements*, called these shapes "Platonic solids," after the Greek philosopher Plato, but Plato is not credited with their discovery.

EXPERIMENT
Using nets

Nets are often used in industry as patterns for pieces of equipment, such as boxes made from flat sheets of cardboard or metal. The net of a polyhedron is the flat shape made when the polyhedron is unfolded. There is usually more than one possible net for any solid shape. If you undo various cereal packets, for example, you may see that they are put together differently. In this experiment you can make a net for a cuboid box (a box that has rectangular faces). The tabs are not part of the mathematical net, but connect one face to another.

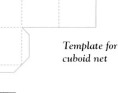

YOU WILL NEED
● *ruler* ● *scissors* ● *pen*
● *double-sided tape* ● *poster board*

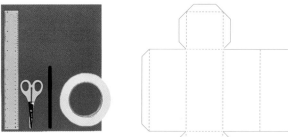

Template for cuboid net

1 USE THE TEMPLATE to make the net for the cuboid. You can either photocopy it until it fits your poster board or use your ruler to measure and scale up the shape. Cut out the shape.

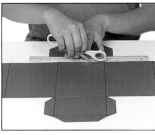

2 SCORE the fold lines (the broken lines on the template) with the blunt edge of the scissors.

3 FOLD the net along the fold lines. Stick the double-sided tape along the tabs to hold the joins together. Assemble the cuboid.

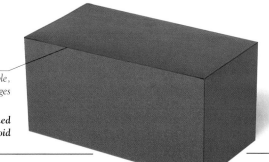

The tabs should be invisible, leaving clean edges

The finished cuboid

EXPERIMENT
Envelope tetrahedron

You can have a lot of fun with solid shapes. In this experiment you make a regular tetrahedron with just an ordinary envelope. A tetrahedron is a pyramid with a triangular base. Pyramids can also have other shapes at their bases, such as squares (like the Pyramids at Gizeh, in Egypt) and hexagons. Some regular tetrahedra fit together exactly. You may like to make more than one of these and join them together to form larger and more intricate shapes. One example is a regular hexahedron — a solid shape that has six faces and is formed from two tetrahedra.

YOU WILL NEED
- *business envelope*
- *scissors* ● *pen*
- *adhesive tape*

1 SEAL THE ENVELOPE with adhesive tape. Fold it in half lengthwise, crease it, then unfold it. Fold up one corner until it touches the center fold. Make a mark at this point.

2 UNFOLD AGAIN. Working from the same corner, take the long edge and fold through the mark made earlier. Unfold. Take the other long edge and fold that through the mark.

3 FOLD THE envelope crosswise through the point, and make a crease. Unfold. Sharpen all the creases with your fingers. (The folds should already show the shape of a triangle.)

4 CUT A LINE 1 in (2.5 cm) from the last crease that you made. The cut should run parallel with this crease, opposite the triangle of folds.

5 USE the scissors to slit open the envelope at each side of the open end, stopping at the horizontal crease through the mark.

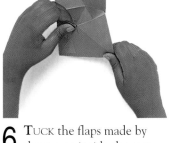

6 TUCK the flaps made by these cuts inside the rest of the envelope. Smooth them flat inside, then sharpen the folds with your fingernails.

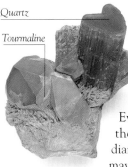

7 OPEN OUT the envelope. Press the sides so that the open ends come together to form the tetrahedron. Hold the corners in the other hand.

8 MAKE SURE that all the sides of the tetrahedron are straight, then stick clear adhesive tape across the open edges to join them together.

The finished tetrahedron

▦ Crystal structure

Quartz

Tourmaline

Crystals, such as those of quartz and tourmaline, are formed with regular arrangements of atoms. Minerals are often identified purely by the mathematical arrangement of their crystals, which is called the crystal system. Even if two minerals are chemically the same, such as graphite and diamond (two forms of carbon), they may have different crystal systems.

Cubes, pyramids, and spheres

THREE OF THE MOST BASIC SOLID SHAPES are cubes, pyramids, and spheres. An understanding of these shapes is crucial in many fields of science and engineering; for example, it can help chemists to predict the likely forms of molecules. The ancient Egyptians used their knowledge of solid shapes for practical purposes such as the construction of the Pyramids. This was an incredible feat of engineering, given the rudimentary mathematics — and tools of construction — at that time. The Great Pyramid of Khufu at Gizeh, for example, measures 760 ft (230 m) along each side and is about 480 ft (146 m) high. Each surface had to slope at the correct angle so that all four sides could meet at the top. The Greek mathematicians dramatically advanced the study of solid geometry, finding new relationships between shapes and developing mathematical proofs. Euclid (p. 114) put forward many propositions on the geometry of solid shapes, and much of his work is still in use.

Recording the Pyramids

This is a page from the Rhind Papyrus (p. 14), written in about 1650 B.C. Problem 56 of the Rhind Papyrus makes special reference to simple trigonometry (pp. 126–127) in the construction of the Pyramids. We know that the Egyptians calculated the volume of a Pyramid as ⅓ × area of base × height, although mathematically they had no way of proving that this was true for all square-based pyramids everywhere.

EXPERIMENT
Make a cube

You can make one shape out of a collection of smaller shapes. In this experiment several irregular pyramids are put together to form a perfect cube. Many puzzles are put together in this way. See what other shapes the pyramids will form. All the pyramids here have square bases. Some pyramids have triangles on all sides (p. 153).

YOU WILL NEED
● *ruler* ● *pen* ● *scissors* ● *double-sided adhesive tape* ● *poster board in 3 colors*

Template for net of square-based pyramid

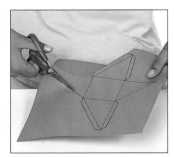

1 COPY THE TEMPLATE for the net onto each piece of poster board by drawing it to scale, photocopying it and transfering it onto the board, or enlarging it with a pantograph (p. 157). Cut out the nets.

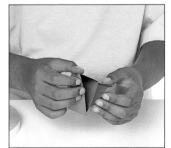

2 SCORE ALONG the dotted lines with the scissors, and bend back all the tabs. Fold in the triangles. The pyramid should hold its shape by itself, but you can fix the tabs with double-sided adhesive tape.

3 PUT TOGETHER two more pyramids in the same way that you made the first. Then see if you can fit the three shapes together so that they form a cube. (The assembly is not as easy as it seems.)

Cube formed from pyramids

EXPERIMENT
Using a cylinder to investigate a sphere

In this experiment you calculate the surface area of a sphere, such as a tennis ball, without having to use the formula (which is $4\pi r^2$). An open-ended cylinder with the same diameter and height as a sphere will have the same surface area as that sphere. Therefore, just measure the base and height of the rectangle used to form the cylinder (p. 98), and multiply these numbers together to find the curved area of the cylinder.

YOU WILL NEED
- *calipers* • *ruler*
- *pen* • *scissors*
- *tennis ball*
- *poster board*

1 FIND the widest point of the tennis ball. Hold the ball still, and use the calipers (p. 95) to measure the diameter. It may help to make a note of this figure.

2 CUT OUT a strip of poster board. The width should be the same as the diameter of the ball, and the strip should be long enough to wrap around the ball at least once.

3 WRAP the poster board strip around the ball, keeping the edges aligned and the sides straight. Mark where the end meets the main part. Cut the strip to that length.

4 MEASURE the base and height of the poster board rectangle. Work out the area. Find the area of the ball using the formula, to see if it matches your figure.

EXPERIMENT
Making molecular structures

A molecule, the smallest component of a chemical compound, consists of atoms linked by chemical bonds. The shape of a molecule governs its chemical, physical, and, in some cases, biological properties. Many simple molecules have their atoms at the vertices of familiar shapes. In this experiment, you can make models of a water molecule (isosceles triangle) and a methane molecule (regular tetrahedron).

YOU WILL NEED
- *colored table tennis balls* • *pair of compasses*
- *toothpicks*

1 TO MAKE the water molecule, choose one ball to represent oxygen. Using the compasses, make a small hole in the oxygen ball.

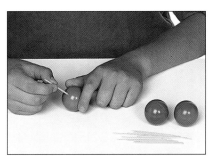

2 INSERT a toothpick into the hole in the ball, making sure that it fits tightly. Next, choose two balls of another color to represent the atoms of hydrogen.

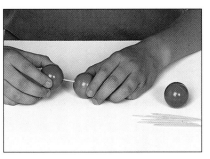

3 MAKE A hole in one hydrogen ball, and fit it on the other end of the toothpick. Attach the other ball to the one for the oxygen, to form a wide isosceles triangle shape (p. 122).

H_2O *(water)*

CH_4 *(methane gas)*

155

Transformations

TRANSFORMATION GEOMETRY shows how shapes change position and size according to certain rules. Some of the most common mathematical transformations are reflections (flips), rotations (turns), translations (sliding without turning), and enlargements or reductions (making larger or smaller). A geometrical shape can be reflected (recreated in its mirror image) in two or three dimensions. It can be rotated by being turned around a chosen point inside or outside it. When a shape is translated, it is moved in any defined direction while its form stays the same. Enlargement or reduction involves multiplying or dividing a shape's dimensions by a specific amount; one example is magnification (p. 22). Transformations are used by scientists to analyze diverse phenomena, from the way in which sand dunes alter over time to the formation of a frog.

(p. 22)

■ DISCOVERY ■
Leonardo da Vinci

An Italian artist and engineer, Leonardo da Vinci (1452–1519) was one of the greatest geniuses of the Renaissance. He created two of the best-known paintings of all time — the *Mona Lisa* and *The Last Supper* — and filled many notebooks with his images of scientific discoveries and innovations. He drew and wrote with his left hand, and used mirror writing to record his work. Although unusual, it was easy to read in a mirror. This illustration shows his design for a flying machine, with ladders that can be pulled up during flight.

EXPERIMENT
Deflation using a balloon

Inflation and deflation are transformations in three dimensions. They can be irregular, as this experiment shows. The appearance of the deflated image will depend on the size and shape of the balloon. You can achieve amusing changes with this method. Show the deflated balloon to friends. Can they guess what the original shape is?

YOU WILL NEED
● *2 magic marker pens* ● *balloon*

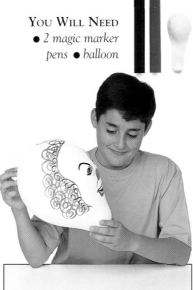

1 INFLATE THE BALLOON and draw a picture on it. This picture shows a clown's face, but you could draw anything, or perhaps write a word in large letters.

2 PRICK THE BALLOON near the knot, or undo the neck, to let the air out slowly. (Take care not to burst the balloon.) As it deflates, the image will be transformed.

DEMONSTRATION
Mirror writing

In mathematical terms, mirror writing is a reflection of the normal letter shapes. Try to write a word so that it can be read when reflected in a mirror.

Tilt the mirror upward slightly, so that you can see your hand and the pen as you write

How to do mirror writing
Prop a mirror on a table with some modeling clay. Watch your hand, and try to form each letter so that it looks correct in the mirror. To write upside down, put the mirror above the paper, as shown here. To write back to front, put it to one side of the paper.

EXPERIMENT
Changing the size

A pantograph is an instrument used by designers to produce technical drawings. This experiment shows how you can make your own simple pantograph, which you can then use to create enlarged, reduced, or inverted drawings of any shape. The instrument is turned upside down, with the heads of the bolts touching the paper, and is fixed to the paper with a thumbtack. You go over the lines of a shape with a pointed piece of dowel, and a pencil attached to the pantograph will automatically produce a transformed image.

YOU WILL NEED
- *drill and bit*
- *bradawl* • *pencil*
- *pen* • *craft knife* • *nuts*
- *bolts* • *large-headed thumbtack* • *washers*
- *cork* • *pencil sharpener*
- *sandpaper* • *dowels*
- *ruler* • *balsa wood*
- *cutting mat*

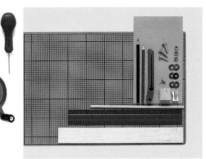

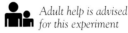 *Adult help is advised for this experiment*

1 ASK an adult to help you cut the pieces of balsa wood. You will need two 8¾×¼ in (22×2 cm) pieces, and two 4¾×¼ in (12×2 cm) pieces.

2 USING a piece of sandpaper, smooth the ends of each balsa wood piece into a gentle curve.

3 MEASURE FIVE EQUALLY spaced points along each piece (one just in from either end) and mark ×'s where the holes are to be drilled. Ask an adult to drill a hole through each mark.

4 JOIN ONE SMALL ARM to each long arm halfway along. Push a bolt through, then thread a washer onto it and fix in place with a nut.

The pencil point must be long and thin enough to reach the paper

5 MAKE A HOLE through the center of the cork. Push the pencil through it, then through the end holes of the small arms.

6 SHARPEN A DOWEL with the pencil sharpener. This is your pointer tool. Insert the dowel through the hole at the end of the right arm.

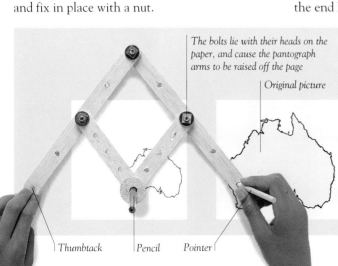

The bolts lie with their heads on the paper, and cause the pantograph arms to be raised off the page

Original picture

Thumbtack   *Pencil*   *Pointer*

7 PLACE the pointer on the image. Lay a piece of paper under the pencil. Put the thumbtack into the left arm. Hold it with one hand and trace the shape of the original picture with the pointer. What sort of image does the pencil produce?

Thumbtack   *Original picture*   *Pencil*

Making Australia larger
Artists who specialize in miniature paintings use pantographs to give greater detail to their work. The pantograph is made of parallelograms, and the fixed point determines whether an image is enlarged, reduced, or inverted. Here, a map of Australia is being enlarged.

Symmetry

MANY NATURAL AND MAN-MADE OBJECTS are symmetrical in two or three dimensions. Symmetry can be rotational, when the image is the same all the way around a central point, or lateral, when the image is perfectly reflected on either side of a line or axis. Many geometrical shapes, such as a circle or a square, are perfectly symmetrical, as are various natural forms. Raindrops falling through the air are symmetrical spheres; honeycombs are symmetrical hexagons; viruses are often three-dimensionally symmetrical shapes. Most manufactured items, such as furniture, airplanes, or highways, have rotational or lateral symmetry. This is partly to suit the nature of the world we live in — chair legs of different shapes would make a chair unstable, and lanes of varying widths would confuse drivers — and partly due to the symmetrical nature of the machinery used to manufacture them.

Common Starfish

This type of starfish has five arms. Its symmetry is lateral, reflected on either side of a line through one arm, and rotational, in that each arm around the central point is the same. This starfish belongs to a class of animals that also includes sea urchins. They all have rotational symmetry of order 5 (symmetry in five positions).

EXPERIMENT
Identical figures

You can make a chain of symmetrical figures from folded paper, by using the mathematical property of reflection. The figures are reflected on either side of the center line when the paper is unfolded. When you draw the outline, make sure that at least one part of it extends over the fold, otherwise the paper will not make a chain. Use decorative holes so that the chain looks interesting when unfolded. Here, a line of doll shapes has been made.

YOU WILL NEED
- *pen* • *scissors*
- *paper*

Make sure some lines extend across the full width

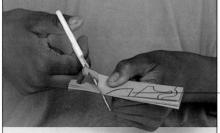

1 FOLD the paper with 1-in (2.5-cm) zigzag folds. Draw half of your design up to the central folds. Some lines should lie across the other folds and the raw edges, to make the links for the chain. Cut out through all layers.

2 OPEN the paper. The line of shapes will be linked, and each shape will be identical on both sides of the central folds.

Details make the shapes more interesting

DEMONSTRATION
Asymmetry

Asymmetry means "absence of symmetry." Do you think your face is symmetrical? After all, you have two eyes, two cheeks, and two ears. Look at the picture on the right. The image has been split in two, and each half has been reflected on either side of the central line to give the pictures below. On the outside our bodies are almost symmetrical, but inside they are not. For instance, we have only one heart, on our left side.

Original image

Reflected symmetry of the left side

Reflected symmetry of the right side

EXPERIMENT
Making a flying machine

A helicopter is an aircraft with rotor blades that enable it to be maneuvered in the air more easily than any airplane. The blades are shaped like the aerofoils on an airplane (p. 107) to provide lift. The rotor is horizontal when the helicopter is ascending or hovering, and is tilted at varying angles for flying in different directions. This experiment shows how you can make your own "helicopter" by using rotational symmetry. This will ensure that the blades are all exactly the same.

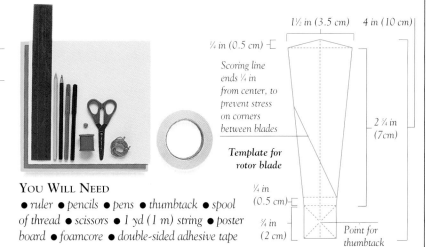

YOU WILL NEED
● *ruler* ● *pencils* ● *pens* ● *thumbtack* ● *spool of thread* ● *scissors* ● *1 yd (1 m) string* ● *poster board* ● *foamcore* ● *double-sided adhesive tape*

1½ in (3.5 cm)   4 in (10 cm)

¼ in (0.5 cm)

Scoring line ends ¼ in from center, to prevent stress on corners between blades

2 ¼ in (7cm)

Template for rotor blade

¼ in (0.5 cm)

¾ in (2 cm)

Point for thumbtack

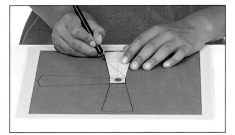

1 TO DRAW the blade shape on poster board, either photocopy the template, enlarging by 200% and then a further 15%, or draw it following the dimensions on the template. Cut out the shape.

2 LAY THE poster board on the foamcore. Pin the template to the poster board and foamcore at its central point. Trace around the shape once. Turn the template 90° clockwise, and trace the shape again.

3 WHEN YOU have traced four blades, draw the scoring line on each blade. Cut out the rotor shape and use the blunt edge of the scissors to score along the scoring line on all four blades.

All the blades slant at an equal angle for level flight

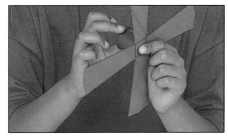

4 USE DOUBLE-SIDED tape to secure the spool of thread beneath the center of the rotor. The score lines should be on the top. Bend along the lines to make the blades slant downward at an equal angle.

5 HOLD the string against the spool of thread and wind it evenly around the sides, starting at the base or the top. It should go around about 14 times.

6 PUT THE BLUNT end of a pencil just inside the central hole in the spool of thread. Hold the "helicopter" upright, with the spool balanced lightly on your thumb and index finger.

7 HOLD the pencil in one hand, and with the helicopter above your head, pull the string quickly and steadily, until it comes off the spool of thread. The helicopter will fly off the end. Do not jerk at the string; the movement needs to be smooth for lift-off.

When the string comes away from the spool, the "helicopter" will take off

From two to three dimensions

PROJECTIONS OF A three-dimensional object can be made to convert the object into a two-dimensional image. If you shine a light on a chair so that a silhouette appears on a wall behind it, you are projecting the chair in two dimensions onto the wall. But you can also take three two-dimensional projections of the chair — one from the side, one from above, and one from the front (p. 162) — to reconstruct it in three dimensions. This is how engineers and architects make plans of structures and equipment. Projections from two to three dimensions may be used for other purposes. For example, forensic scientists and archeologists recreate faces from pictures and skulls, and modelmakers in waxworks museums take photographs and measurements of their subjects when making three-dimensional models. Map projections are two-dimensional but depict a near-spherical Earth by using lines of longitude and latitude to define the positions of places or areas on it.

▦ Mercator projection

The Flemish geographer Gerardus Mercator (1512–94) developed this form of map in the 16th century. His projection is made from the Earth's center onto an imaginary cylinder around the equator. Navigators use it to plot a course by drawing straight lines on a map. However, it distorts areas near the poles, making them seem disproportionately large. Shown here is Mercator's map of the Western hemisphere, made in 1631.

EXPERIMENT
Traveling in a straight line

Mercator projections were the first modern type of map projection. They provide directional information, which enables navigators to plot courses for ships and aircraft. Follow this experiment to see how a route between two cities can be planned. First, the route is shown on a globe, then the information is projected onto a map of the world. You may expect the most direct route between two locations to be a straight line on the map, but are you right?

YOU WILL NEED
● *pencil* ● *pens*
● *scissors* ● *string*
● *modeling clay*
● *adhesive tape*
● *tracing paper*
● *world atlas*
● *globe*

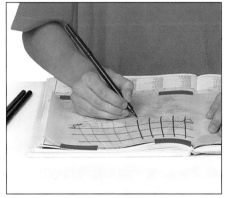

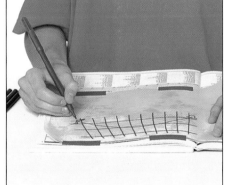

1 PLACE LUMPS of modeling clay on two cities on your globe. Cut a piece of string slightly longer than the distance between the lumps. Pull it taut, then push the ends into the lumps, to show the most direct route on the surface of the globe between the two cities.

2 FIX A PIECE of tracing paper over the world map in an atlas. Use adhesive tape to hold the paper steady. Draw on it the latitude (east–west) and longitude (north–south) lines that cover the area crossed by the string on your globe, and go over the lines in pen.

3 PLOT THE ROUTE on the tracing paper. Look at the points where the string touches the latitude and longitude lines on the globe. Mark these coordinates onto the tracing paper laid over your map. Join up the marks. What do you discover about your route? (Answer on p. 187.)

EXPERIMENT
Making a face mask

Using pictures and measurements in two dimensions, you can make a clay mask of your own face. Measure the length of your face. Measure its width at the points shown on the picture below to help you find the position of your eyes, nose, and mouth. Your profile measurements, taken from a line just in front of your ear, will help you shape the contours of your face.

YOU WILL NEED
- apron ● clay for sculpting ● hand drill
- tape measure ● ruler ● pens ● wooden skewers ● set square
- photographs of your profile and full face
- paper ● 8×12 in (20×30 cm) balsa wood board ● scissors

Adult help is advised for this experiment

Measure the width of your face and your features

Full face

2 PHOTOCOPY your full-face picture, enlarging it to the same size as your face. Trace the features onto paper. Write the measurements on your drawing, then copy it onto the wood.

Profile

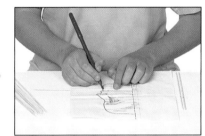

3 PHOTOCOPY your profile picture and trace it onto paper, as above. To reproduce its dimensions on the board, lay a skewer on each line, mark the length, then cut the stick to that length.

1 ASK A FRIEND to measure your face and note the dimensions. Measure down a center line through your nose, then at vertical and horizontal intervals of about 1½ in (4 cm), including all the widths shown in the pictures on the right.

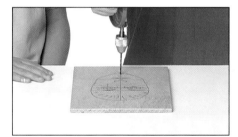

4 ASK AN ADULT to drill a hole in the board at each point where you took a profile measurement, such as the tip of the nose and the edges of the eyes. The skewers should fit securely in the holes.

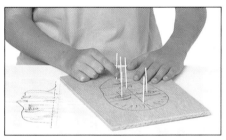

5 INSERT A SKEWER of the correct length at each point to show the depth of your face at these places. The balsa-wood picture is now ready to form the base for your clay model.

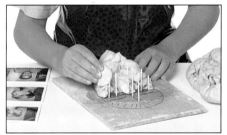

6 BUILD UP THE CLAY around the skewers, until it just covers the tips. Keep it damp, to make your work easier. Sculpt your features roughly, using the photographs as a guide.

▦ The sum of the angles

The interior angles of a triangle add up to 180° (p. 122). However, in non-Euclidean geometry (p. 164), this is not always so. Draw a triangle on a hollow, transparent sphere such as a fish bowl. Measure the angles on the outside with a flexible protractor, which you can make by photocopying a plastic one onto tracing paper. Then measure the triangle from inside the bowl. What do you find? (Answer on p. 187.)

7 SHAPE the features with more detail. You can use the tips of the scissors to draw your hair and lips. Wrap the finished mask in a damp cloth to prevent the clay from drying out and becoming brittle.

Projections in art

PROJECTIONS HAVE BEEN used in art for hundreds of years. In 1435 the artist Leon Battista Alberti (1404–72), in his book *On Painting*, used the word "composition" to describe the way geometry could be used in art. Many 15th-century artists like him were also mathematicians, and statesmen gave them problems of construction to solve. These artists used the mathematics of perspective to design churches, palaces, and other buildings, and to create paintings that were as realistic and three-dimensional as possible. Today, computer-generated plans of buildings incorporate all the necessary information to construct them. Some computers can use the data from these plans to create a virtual building around which the designer or the potential user can walk and visualize the final structure.

The Grand Canal, Venice

The Italian painter Antonio Canaletto (1697–1768), whose works include many paintings of Venice, started work as a painter in the theater. This work required the use of perspective. In the 1730's he developed a device like a camera obscura (p. 114) in which a lens threw an image on to a glass screen. He then used this image as the basis for a drawing. People and buildings were placed in his paintings according to a strict formula.

EXPERIMENT

Isometric drawing

Isometric drawings are used by technical illustrators and engineers to create three-dimensional images of objects, with undistorted measurements of the objects' most important dimensions. Isometric paper (available at an art store) is marked with a triangular pattern of lines at 60° to each other. This experiment shows how you can draw a three-dimensional numeral by using this paper. Start by drawing the front view (the front elevation), then draw the view from the end (the end elevation) and finally the top view (the plan).

YOU WILL NEED
● *ruler* ● *pen* ● *graph paper*
● *isometric paper (the paper used here has ¼-in/5-mm sided equilateral triangles)*

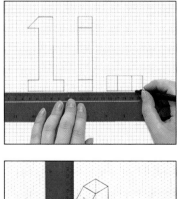

1 DRAW THE PLAN, and the front and side elevations, of the number 1 on the graph paper. You can make up your own measurements for the 1, but they must be consistent on all the views. Here, the base is three squares high and four deep.

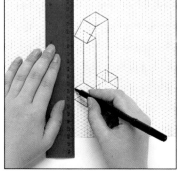

2 CHOOSE ONE POINT on the front elevation, say the bottom left-hand corner. Measure the length of the numeral's base, and draw a line of that length on the isometric paper. Your line will be at an angle of 30°. The paper naturally guides you to draw in three dimensions.

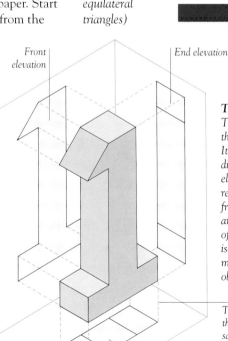

Front elevation

End elevation

Plan

The drawing
This illustration shows the finished number 1. It appears to be in three dimensions, with the elevations and the plan represented by the views from the front, the side, and the top. Because of the nature of isometric drawings, the measurements of the object remain accurate.

The measurements on the finished shape are the same as the corresponding ones on the flat views

EXPERIMENT
Drawing with a screen

The German artist Albrecht Dürer (1471–1528) stressed the importance of precision when using perspective. He sometimes used a screen similar to the one below. By copying the outline of his subject square by square through the screen, he could draw it with the correct perspective. Make your own Dürer screen, and draw some objects in perspective.

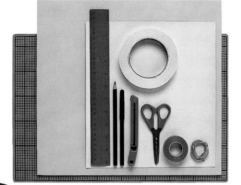

YOU WILL NEED
● *ruler* ● *double-sided adhesive tape*
● *pencil* ● *pen* ● *craft knife* ● *scissors*
● *adhesive tape* ● *string* ● *paper*
● *2 pieces of foamcore* ● *cutting mat*
● *objects for drawing (such as fruit)*

Adult help is advised for this experiment

1 Draw a 16-in (40-cm) square on foamcore and measure a 2-in (5-cm) wide frame inside it. Draw a strip 2 in (5 cm) wide along one outer edge of the frame, to make the base. Ask an adult to cut out the center with the craft knife. Make another identical frame.

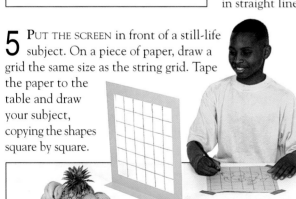

On the back of the frame, score along the line at the base

2 Turn over one of the foamcore frames. With a ruler and a pen, mark the inner edges of this frame on all four sides, at 2-in (5-cm) intervals, to indicate where you will attach the pieces of string for the grid.

3 Use small pieces of adhesive tape to attach the strings across the frame at the marked points. Make sure that you pull all the strings taut so that they lie in straight lines.

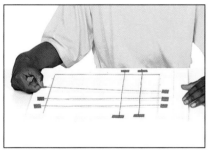

4 Put double-sided tape around the square part of the second frame, to fix it to the back of the first one. Cover all the tape and string ends. Bend the base strips out to form the stand.

5 Put the screen in front of a still-life subject. On a piece of paper, draw a grid the same size as the string grid. Tape the paper to the table and draw your subject, copying the shapes square by square.

What the artist sees
This is how an artist would see the grid lines through the screen, merged with the object that is being drawn. After the artist has copied the outline of each shape in the picture, using the screen, his or her skill must then be applied to turn this outline sketch into a finished work of art.

What is topology?

TOPOLOGY IS SOMETIMES called rubber-sheet geometry. All other geometry is the study of rigid shapes and their angles, lengths, and curvature. But topology deals with a collection of physical points that can be deformed. Total flexibility is the rule. In topology, a square and a circle, for example, are said to be equivalent because one can be stretched and squeezed until it looks like the other, without any holes being closed up or any new ones being created. Measurements such as the length of a side do not matter in this area of mathematics. A handful of modeling clay can be transformed topologically (p. 104). Knot theory (p. 167) in topology is used by scientists to solve complicated questions about how molecules arrange themselves. Network theory, which is often used in the design of electrical circuits, is another practical application.

INVESTIGATING SHAPES
Euler's theorem

Leonhard Euler (1701–83) was an important mathematician who stated that in any polyhedron there is a relationship between the numbers of vertices, edges, and faces (p. 152). He stated that $V - E + F = 2$, where V is the number of vertices, E is the number of edges, and F the number of faces.

OCTAHEDRON
This shape has 6 vertices, 12 edges, and 8 faces. If Euler's formula is applied to it, the result is $6 - 12 + 8$, which equals 2.

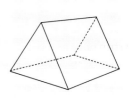
Vertex *Edge* *Face*

PRISM
A triangular prism has 6 vertices, 9 edges, and 5 faces. Applying the formula gives $6 - 9 + 5$, which also equals 2.

IRREGULAR SHAPE
This polyhedron has 24 vertices, 48 edges, and 26 faces. Applying the formula gives $24 - 48 + 26$, which equals 2.

■ DISCOVERY ■
Karl Friedrich Gauss

Gauss (1776–1855), a German mathematician, invented complex numbers (such as the square root of a negative number) and contributed to the science of statistics. His most revolutionary work dealt with non-Euclidean geometry. In traditional Euclidean geometry (p. 117), which deals with objects on flat surfaces or in ordinary space, parallel lines never meet. By contrast, in non-Euclidean geometry, parallel lines can meet or even cross.

DEMONSTRATION
Undressing trick

How can someone wearing a woolen vest underneath a sweater take off the vest without removing the sweater? Look carefully at the steps shown on the right, to see if you can understand how the process works, then try it yourself. The number of holes and surfaces on the two garments is fixed. The relationship between these holes and surfaces is used to solve the puzzle. This process is similar to the solution for the rope puzzle (see opposite page), which also uses topology.

1 PUT ON the woolen vest, then put on the sweater over the vest.

2 PULL the right armhole of the vest wide. Put your right elbow, then arm, through it from the outside.

3 PULL the vest across your back toward your left arm. Try not to get the two garments tangled up.

EXPERIMENT
Cutting a Möbius strip

This mathematical wonder is made from a flat strip that is given one twist, then joined to make a loop with just one side and one edge. Chemists use mathematics to create a range of molecular Möbius strips with unusual characteristics. Make your own Möbius strips, and see what happens when you cut along the strips in various ways. (Answers on p. 187.)

YOU WILL NEED
● *pen* ● *scissors* ● *adhesive tape* ● *double-sided adhesive tape* ● *paper in 2 colors*

1 FROM EACH piece of colored paper, cut a strip 24×2 in (60× 5 cm) long. Stick one strip over the other with double-sided tape. Twist the strip once and join the ends.

2 DRAW A LINE along the middle of the strip. Cut along this line from the joint, until you return to your starting point. What happens to the strip when you finish?

3 DRAW ANOTHER line along the center of the cut strip. Cut along this line from the joint. What has happened to the Möbius strip by the time you reach the joint again?

Cutting in thirds
Make another Möbius strip. This time, cut along the strip only a third of the way in from one edge. What happens to the strip when you finish cutting?

Klein bottle
Felix Klein (1849–1925) was fascinated by non-Euclidean geometry. He developed theories about surfaces and shapes, and invented the Klein bottle — a three-dimensional tube whose inside surface loops back on itself to merge with the outside. The Klein bottle, visible only when generated on a computer (right), not in three-dimensional space, has interesting topological properties. The bottle is one-sided (like the Möbius strip), is a closed shape, has no ends, and yet has no inside. If cut in two lengthwise, it would form two Möbius strips.

123 Trick
Try this with a friend. You need two lengths of string or rope. Tie the ends of one rope to your friend's wrists. Tie one end of the other rope around your left wrist. Pass the free end under the rope between your friend's wrists, and tie

it to your right wrist. You should be able to unlink your rope from the one on your friend's wrists, without untying the ends from your wrists. Can you see how to do this? (Answer on p. 187.)

4 PULL the vest around to the front. Stretch the right armhole again. Put your left arm through, elbow first.

5 YOUR LEFT ARM is still in the vest. To free it, push the vest down the left sleeve of the sweater.

6 WORK the vest all the way down the left sleeve of the sweater, and pull it out the end.

7 YOU have now managed to remove the vest completely without removing the sweater on top.

Practical puzzles

TOPOLOGY WAS INITIALLY THOUGHT OF as a form of geometry, but many other uses have been found over the years, such as algebraic topology (which solves topological problems using algebra) and certain kinds of number theory. The tying of knots is another part of topology. Knots are complete curves that snake through three-dimensional space. Untying a complex knot can involve mathematics. For instance, given a mass of knotted string, a mathematician could work out whether or not it could be transformed into a simple knot. This exercise helps scientists understand the behavior of DNA molecules (p. 149), using knot theory to determine whether pairs or small groups of molecules are tied together in similar ways.

Puzzle

The "map" shown below is a closed curve that crosses itself. Try to color it so that shapes with a shared boundary have different colors. What is the minimum number of colors you need? (Answer on p. 187.) Mapmakers face a similar problem, which was finally solved by a computer (p. 187). Trace a map, and try it yourself. Do you agree with the computer?

EXPERIMENT
Making a magic shape

In this shape, cuts and folds are used to arrange the apparently random numbers on a grid into groups of the same number. Begin by drawing a grid of four by three squares, making the sides of the squares about 1 in (2.5 cm). Fill in the numbers on both sides of the grid, following the templates (right).

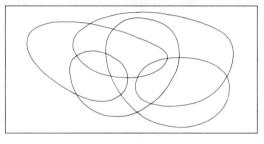

4	4	3	2
2	3	4	4
4	4	3	2

Front view

1	1	2	3
3	2	1	1
1	1	2	3

Back view

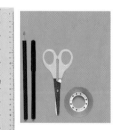

YOU WILL NEED
● *ruler* ● *pencil* ● *pen*
● *scissors* ● *clear adhesive tape* ● *paper*

1 WHEN YOU have filled in the numbers, cut out the grid. Cut along three sides of the middle two squares with the scissors, as shown.

2 TURN the paper over and fold in the left-hand side. Open out the center flap and fold it back. Fold in the left-hand side again.

3 YOU SHOULD NOW have a rectangle with all the 2's together. Fold the center flap around the right-hand edge of the rectangle.

4 TURN the shape over so that the side with all the 1's faces up. Using clear adhesive tape, join the two central 1's together.

5 TURN the magic shape over again so that all the 2's are facing upward. Open out the central hinge and see what happens.

6 THE 3's appear as if from nowhere. Can your friends figure out how you did this?

EXPERIMENT
Plug for different holes

Sometimes a single shape can be used to give the solutions to more than one visual problem. For example, in the following experiment you can mold modelling clay to build one plug that will fit equally well into three different plug holes. The experiment encourages you to think about non-regular shapes.

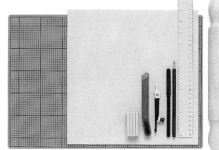

YOU WILL NEED
● *modeling clay* ● *craft knife* ● *pair of compasses* ● *pen* ● *pencil* ● *ruler* ● *foamcore* ● *cutting mat* ● *rolling pin*

Adult help is advised for this experiment

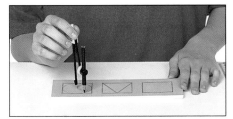

1 CUT A piece of foamcore 8×3 in (20×8 cm). Draw three 2-in (5-cm) squares evenly spaced on it. Draw a circle in one square and a triangle in another.

2 SOFTEN THE modeling clay and roll out six lumps to about ¼ in (6 mm) thick. They should be slightly larger than the shapes in the squares.

3 ASK AN ADULT to cut out the shapes from the foamcore. Lay the foamcore over the clay and cut out a circle, two squares, and two triangles.

4 STARTING WITH the circle as a base, use a square to build one side of your plug. You may find it useful to put some clay inside the shape to hold it.

5 PUT ANOTHER square opposite the first. Squeeze the tops together, then place the two triangles on the open sides, curving the bottom edges to fit.

6 SMOOTH ALL the rough edges of the clay with your hands, but without losing the shape of the pieces forming the sides. Then position the sides of your shape against each of the holes that you cut in the foamcore strip. Does the plug fit all three holes equally well?

EXPERIMENT
Transferring loops

Knots and loops are studied in topology. In this experiment a loop can be transferred from a piece of paper to a pair of paper clips. Can you do it with more paper, more loops, and more paper clips?

YOU WILL NEED
● *ruler* ● *pencil* ● *2 paper clips* ● *scissors* ● *paper*

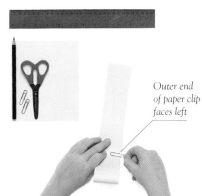

Outer end of paper clip faces left

1 CUT A STRIP of paper about 2 in (5 cm) wide. Fold one end under to make a loop. Secure it with a paper clip as shown above.

2 FOLD OVER the other end of the paper to make a second loop. Clip it to one side of the loop that you made in step 1. The outer end of the paper clip should face left.

3 TAKE HOLD of both ends of the paper strip. Pull the ends apart sharply, so that the loops vanish and the strip straightens out. What happens to the two paper clips?

THINKING

Puzzling over the paths
Markings on an early gravestone (above) in Alkborough cemetery, northern England, resemble the pattern of a maze, a popular puzzle for many hundreds of years. A beautiful image of fractal geometry (left) is built up on a computer program from the Mandlebrot set (p. 180) and is used to analyze certain chaotic systems.

PUZZLES that test people's skill with logic have always been popular. Games such as chess are also founded on elaborate methods of reasoning. However, progress in improving our thinking skills came about only when scholars began to analyze thought processes in a mathematical way. In the last 200 years logic has been developed into a branch of mathematics, with its own symbols and methods, and today finds applications in computers and their programs. More recently, new ways of thinking have opened up further areas of mathematics. Chaos theory and fractal geometry have enabled scientists to measure and explore phenomena in the natural world — including ocean waves, weather systems, and protein molecules and their attractions — in a new way.

THE NEW GOLDEN AGE

Rᴇᴀsᴏɴɪɴɢ, ᴛʜᴇ ᴀʙɪʟɪᴛʏ ᴛᴏ ᴛʜɪɴᴋ ᴛʜʀᴏᴜɢʜ ɪᴅᴇᴀs ʟᴏɢɪᴄᴀʟʟʏ, is the cornerstone of mathematics. Reasoning is also used in science, philosophy, and language. In the past, mathematics was regarded as a complete system of perfect logic. However, in the 20th century, gaps have been found in the system and new disciplines have been created to fill these gaps.

We all use our natural powers of reasoning to make sense of the world around us. For example, if we are told that Napoleon and Louis XIV were both Frenchmen and that all Frenchmen are (or were) Europeans, we can work out that Napoleon and Louis XIV were both Europeans.

Mathematicians use reasoning to link chains of ideas in proofs, or to reduce a complex set of equations or ideas into simpler ones. For example, if two shapes called A and B have the same area and two shapes called B and C have the same area, then they can work out that A and C also have the same area. Unlike the reasoning that we use in everyday life, mathematical reasoning is based on a system of symbols and ideas that have precise meanings. Reasoning can be applied to link ideas in mathematics only with a set of special principles called "logic." The starting point of a chain of logically connected ideas is called a "premise," and the conclusion is reached by a process that is called "deduction."

Logic and algebra

We already have an intuitive grasp of many principles of logic. For instance, the example given above, involving Napoleon and Louis XIV, illustrates our natural understanding of "subsets." These are sets of objects fully contained within other ones. In this case, we have the premise that "Frenchmen" is a subset of "Europeans" and that "Napoleon

Aristotle
The ancient Greek philosopher Aristotle founded the study of formal logic. His philosophy was taken as the basis of Christian theology in the 13th century, and still influences Western philosophy today.

Overlapping sets
Venn diagrams show how sets of similar items overlap. Here, one set is of red shapes and the other is of squares. Red squares belong to both sets. Shapes that are not red or square belong to neither.

and Louis XIV" is a subset of "Frenchmen," so we can deduce that "Napoleon and Louis XIV" is a subset of "Europeans."

Mathematicians, however, often apply logic to deduce conclusions from more complex premises that cannot be worked out with intuition. Some conclusions can be deduced more easily using a "Venn diagram" (above). This is a special drawing that can show how sets overlap. Venn diagrams often help simplify relationships between two or more sets.

A special form of algebra has been devised to help mathematicians to work through really complex deductions. It is called Boolean algebra, after its inventor George Boole (1815–64). Using this system, mathematicians can not only write logical ideas, but also analyze them using different techniques. Boolean algebra can also be applied to work out what happens when a series of logical

functions is performed. For example, imagine two sets: the set "B" of all birds and the set "T" of all flying animals. A typical logical function is "NOT." "NOT B" (sometimes written as $\bar{B}$) means the set of things that are not birds. Another function is "AND." "B AND T" is the set of anything that is both a bird and a flying animal. $\bar{B}$ AND T is the set of anything that is not a bird but is a flying animal.

Philosophers

Many of the people who developed logic were also great philosophers. It was the German mathematician Gottfried Leibniz (p. 47) who first tried to put together symbols to represent logical ideas. However, scholars

Kurt Gödel (1906–78)
In 1951, Kurt Gödel won the Einstein Award for Achievement in Natural Sciences. Here, Einstein himself presents the award to Gödel (right).

had been fascinated by logic long before his lifetime. For example, the ancient Greek philosopher Aristotle (384–322 B.C.) employed mathematics and physics to illustrate the power of reasoning.

In the 17th century, René Descartes (p. 74) used reasoning to understand the nature of existence. Having decided that nothing in the physical world should be believed without proof, he concluded that the only thing that he could be sure of was his own existence. Certain that he existed because he could think, Descartes summed up his reasoning with the famous words "I think, therefore I am." In the 20th century, the foundations of mathematical reasoning were completely shaken by two philosophers: Kurt Gödel (1906–78) and Bertrand Russell (p. 173). In 1931, the Austrian-born Gödel showed that mathematics was not a perfect, complete system of thought but one containing statements that could be neither proved nor disproved. Russell wished to show how mathematics can be based on a complete and logical system, but found that the theory of sets used by mathematicians could generate statements called "paradoxes."

Part of egg-dispensing machine
This machine (p. 178) is made for fun, but the logical process used in constructing it has many practical applications — for example, in the construction of large buildings.

Paradox

A paradox is a strange statement that appears to contradict itself. An expression such as "this statement is false" is an example of a paradox. Many paradoxes suggested in the past, such as

Goldfish problem
Sometimes a proposition can remain unproved for centuries, simply because nobody has ever tested it. This was the case in the problem of the goldfish in a bowl (p. 173).

Zeno's paradox concerning the athlete and the tortoise (p. 172), were resolved once people gained a better understanding of science or mathematics. A few, however, are not dependent on how much we know but result from inconsistencies in the logical system that we use.

Russell's Paradox is an example of a paradox that results from the system of logic that we use naturally. This paradox is very hard to explain in mathematical terms, but it can be illustrated with the example of a barber. Imagine a barber who cuts the hair of only people who cannot cut their own hair. Who cuts the barber's hair?

Applying logic

Despite the problems that Gödel and Russell discovered with mathematical logic, this form of logic can still be employed to solve problems in many areas of life, such as those encountered in town planning, circuit building, and computer programming.

Computers often analyze real-life problems using logic. To do this, computer programmers must represent the problems in such a way that a computer program can work with them. For example, in reality, the road plan of a town may be a network that connects different destinations, such as the town hall, the school, and individual houses. However, in a computer program, such a plan may be represented far more simply as a

network of points (called "nodes") linked by interconnecting lines (known as "paths").

Chaos

Until recently, mathematicians thought that some systems in the real world were totally disordered and could never be understood using mathematics, a system based on reason. Such systems are described as "chaotic."

Examples of chaotic systems are the long-term weather patterns of the world and the flickering of our eyelids while we are in deep sleep. Very small changes in the way each one is started bring about massive, unpredictable changes in the way they behave. The tiniest variations in air pressure at one location in the world, for example, may be responsible for a violent storm at a seemingly unconnected location on the other side of the globe.

Mathematicians have now found ways to study chaotic systems and have devised new forms of mathematics that can mimic the systems' behavior. Their studies have shown that "chaotic" systems behave in a way that can be predicted with the right methods. Mathematicians hope to apply this understanding to forecast long-term weather patterns, movements on stock markets, and population growth and decline.

Tower of Hanoi
This is a simple version of a classic logical puzzle (p. 175), where rings are moved from one needle to another but kept in their correct order. The original puzzle had 64 rings, but the principle is the same.

Town planning
Networks need to be worked out when new urban areas are planned, so that utilities such as gas, water, and electricity can be provided and key buildings and roads are linked in the most efficient way possible.

What is logic?

LOGIC IS THE STUDY of ideas and how they are used in argument. In essence, logic governs the form of an argument rather than the accuracy of the facts. Premises (starting ideas) are linked with words such as "if … then," "and," "or," and "it is not the case," and lead to a conclusion. The Greeks, notably Aristotle (384–322 B.C.), were the first to apply logical processes to statements. Centuries later, Leibniz (p. 47) suggested that logic could be organized in terms of a universal language like algebra. In the 19th century, such a language was devised by the British mathematician George Boole (p. 69). Premises and conclusions are represented by algebraic symbols and are linked by other symbols to form a logical argument. When scientists are confronted by a puzzle, they often work out the answer using this system.

INVESTIGATING LOGIC
What is a syllogism?

A syllogism is an argument in three parts. It consists of two premises, on which the argument is based, and one conclusion. In logic there are rules that state whether or not a syllogism is valid.

1 The first premise must have one thing in common with the second premise.
All men are mortal.

2 The second premise must have one thing in common with the first premise.
No gods are mortal.

3 The conclusion must have one thing in common with both premises.
Therefore, no men are gods.

A leap into space – think about it!

If you jump in the air while traveling in a moving train, and the train continues to move forward while you are in mid-air, will you have traveled backward in the train by the time you land? (Answer on p. 187.)

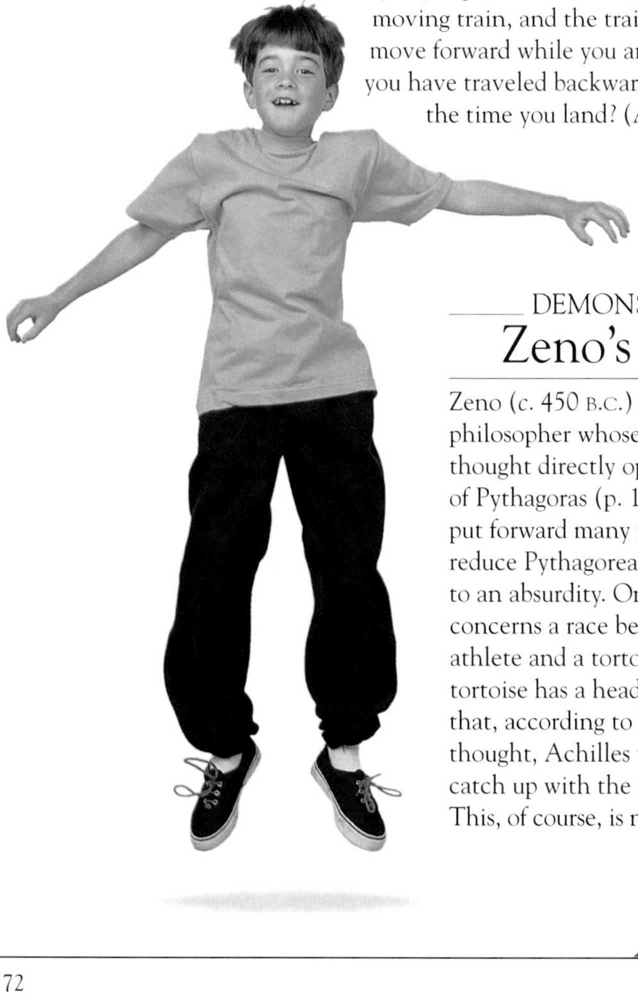

Achilles (shown here on the girl's T-shirt) begins at this point, giving the tortoise a head start

DEMONSTRATION
Zeno's paradox

Zeno (c. 450 B.C.) was a Greek philosopher whose school of thought directly opposed that of Pythagoras (p. 124). Zeno put forward many paradoxes to reduce Pythagorean thinking to an absurdity. One paradox concerns a race between Achilles the athlete and a tortoise, in which the tortoise has a head start. Zeno showed that, according to Pythagorean thought, Achilles would never catch up with the tortoise. This, of course, is nonsense.

Fooling the thinkers

In the 17th century, one of the logical puzzles given to scientists (including some eminent people of that time) was the following: A bowl of water is weighed on a scale. If a fish is added to the water, why does the reading on the scale not increase? Many ingenious and seemingly logical answers were proposed until one day a scientist actually tested the experiment with a fish and a bowl of water, only to find that the bowl did weigh more when the fish was added. So even the thinkers can be fooled sometimes! (See page 102 for the reason why.)

(See page 102 for the reason why.)

■ DISCOVERY ■
Bertrand Russell

Russell (1872–1970) was a British philosopher and logician who also won the Nobel Prize for Literature in 1950. His genius was recognized while he was at Cambridge University, England, where he studied mathematics and philosophy. He aimed to express knowledge in a simplified form, and to link logic directly with mathematics. Russell brought his ideas to a wide audience through writing and broadcasting. He was a pacifist; his activities in World War I caused him to be fired from his lectureship at Trinity College, Cambridge, and imprisoned. He visited the Soviet Union in the 1930's in support of world socialism — a cause which, like pacifism, was unpopular in Britain at the time.

The tortoise (shown on the boy's T-shirt) has a head start, and at a given moment, both runners begin to move forward

By the time Achilles reaches the tortoise's starting point, the tortoise has moved forward to another point

● Grey matter: Paradoxical statements

The most corrected copies are commonly the least correct.

(Francis Bacon)

All animals are equal, but some are more equal than others.

(George Orwell, *Animal Farm*)

I am telling you the truth when I say I am a liar. (Anon)

Less is more. (Anon)

By the time Achilles has covered this distance, the tortoise has moved forward again

By the time Achilles has covered the smaller distance, the tortoise will have moved farther forward, although the distance between them is diminishing

This progress continues with Achilles getting ever closer to the tortoise but never actually reaching and overtaking it

Using reasoning

MATHEMATICAL PUZZLES have existed for thousands of years, but it was only in the 17th century that books of puzzles were first printed. Puzzles can touch on many aspects of mathematics. Some deal with shapes, whereas others require you to solve a problem by using logic. Some, such as the ones on these pages, combine physical and mental challenges. Logical puzzles usually demand no special knowledge of numbers, and often they can be solved only through trial and error. As well as being fun, puzzles can provide important mathematical information. For example, a game involving paths and nodes (p. 177) may reveal new properties of special shapes. Try the puzzles on this page to test your skill in logic.

 Puzzle

If you have a crate divided into 36 spaces, how can you fit 14 bottles into the crate so every row and column has an even number of bottles in it? There are at least three solutions. (Answer on p. 187.)

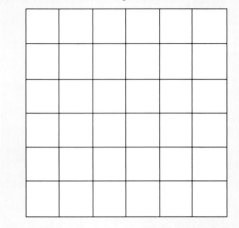

EXPERIMENT
Making paper logic chains

This experiment shows how to put together a sentence in the form of a paper chain. This works in a similar way to a flow chart (p. 179), linking ideas in a logical progression. (The chain, however, does not allow for questions.) It must have marked START and END points, and the links in between must be added in such a way that they form connections that make sense. For example, you could make a chain that shows the types of drinks or foods that you like, using the illustration opposite as a guide. Test your logic by cutting one of the links. Does your chain hold together?

YOU WILL NEED
- *scissors* ● *pens*
- *ruler* ● *stapler*
- *colored paper*

1 CUT OUT 1-in (2.5-cm) wide strips of colored paper. Then write out a plan for the first few links of your logic chain (see illustration, opposite page).

2 MAKE SURE that you have one strip of paper for each item that you wish to include. On the first strip, write "START," then the first part of the sentence.

3 WRITE ONE item on each of the strips. On the last strip, write the last part of the sentence, then "END." Staple the first strip to form a loop to start the chain.

4 ADD LINKS to your chain. The links must always form correct sentences. For example, "I drink coffee dissolved in water, with milk, heated, in a mug . . ."

🧩 Puzzle

There are five regular polyhedra, or "Platonic solids" (p. 152), but there is only one polyhedron whose edges you can trace without going over any edge twice. Which is it? (Answer on p. 187.) Because polyhedra are three-dimensional shapes, you must remember to trace along all the edges, not just the ones at the front.

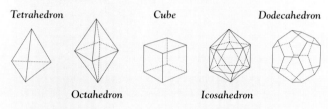

Tetrahedron *Cube* *Dodecahedron*

Octahedron *Icosahedron*

Drinks paper chain

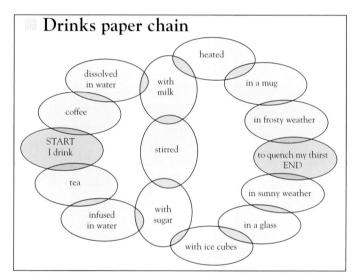

5 WHEN YOU HAVE ADDED all the links to make a complete pattern, ask a friend to hold the chain in the air while you cut one of the links anywhere at random. Follow the new paths that are formed to check if your logic is sound. If the remaining links of the chain form incorrect statements, there may be a mistake in the first parts of your original statement.

The Tower of Hanoi

In the legend of the Tower of Hanoi, a monastery in Hanoi has three needles. One holds 64 gold discs in descending order of size. The monks have orders from God to move all the discs to the third needle while keeping them in descending order. A larger disc must not sit on a smaller one. All the needles can be used. When the monks move the last disc, the world will end. Here is the solution for a version with three rings. Can you work out how many moves are needed if five rings are used? (Answer on p. 187.)

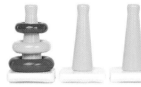

1 ARRANGE the three rings on one cone with the largest ring at the bottom and the smallest at the top.

2 MOVE ONE ring at a time. Start by moving the smallest ring over so that it sits on the far cone.

3 YOU cannot place a ring on top of a smaller ring, so the middle ring must go onto the central cone.

4 NOW move the smallest ring over to the central cone, so that it sits on top of the middle ring.

5 MOVE the largest ring over to the far cone. By doing this, you place the ring in its correct final position.

6 SHIFT the smallest ring back to the first of the cones, so that you leave the middle ring free.

7 PLACE the middle ring on top of the largest ring. You now have two rings in the correct place.

8 PUT the smallest ring on top of the others. You have transferred all the rings in just seven moves.

Logic puzzles

PEOPLE HAVE ALWAYS ENJOYED devising puzzles, either for fun or as an intellectual challenge. As mathematics has developed, so too has the range of puzzles. Mental games now involve geometry, topology, and logic. Puzzles can be grouped into types, and once you have learned to tackle a particular type, other types of puzzles may become simpler to solve.

Puzzle

One interesting logic puzzle involves two glasses, one filled with red-colored water and the other filled with an equal amount of green-colored water. Take one spoonful of green water, and pour it into the glass of red. Then take one spoonful of the red water, and pour it into the green. Is the amount of green water in the red glassful greater than the amount of red in the green glassful, or vice versa? (Answer on p. 187.)

Alice in Wonderland

Lewis Carroll (1832–98), whose real name was Charles Lutwidge Dodgson, was a logician and mathematician who also wrote novels. From a large family, he was accustomed to the company of children. He was also shy, and found children's company friendlier than that of adults. While lecturing in mathematics at Oxford University, England, he wrote a story for a girl called Alice, the daughter of a colleague, and called the story *Alice's Adventures in Wonderland*. The tale contained riddles and logical puzzles that amused adults as well as children. Here, Alice finds a bottle marked DRINK ME, but decides not to do this until she has checked to see that it is not also marked POISON.

The tale of the journey across the stream

A fox, a hen, and a pile of grain need to cross a fast-running stream. The fox and the hen cannot swim across, and the grain, of course, cannot swim at all. A man with a boat is available to take them, but the boat cannot hold more than two as well as the man. The fox cannot travel alone with the hen or stay on the bank with her, because he would eat her. Likewise, the hen cannot be alone with the grain, because she would eat it. How can all three be carried safely to the other side of the river? (Answer on p. 187.)

Hen

Fox

Grain

EXPERIMENT
Building a maze

Mazes consist of paths and nodes. A node is a point at which there is a choice of paths. An "old" node or path is one that you have visited before. A few simple rules can be used to find the center of a maze. Never follow a path more than twice. When you reach a new node, select a path. When you reach an old node or a dead end by a new path, return by the same path. On reaching an old node by an old path, select a new path if possible; if not, select an old path. Build a maze, and ask a friend to solve it following these rules.

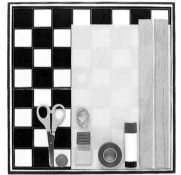

YOU WILL NEED
● *scissors* ● *table tennis ball* ● *string* ● *modeling clay* ● *putty adhesive* ● *adhesive tape* ● *glue* ● *2 pieces of wooden molding* ● *paper* ● *checkered board*

The traditional maze

Mazes have provided fun for people for hundreds of years. During the Renaissance, intricate mazes made with hedges became popular in formal gardens. The hedges were high and thick, so that people inside the maze could not see from one path to another. The gardens of many great European houses, including Longleat House in England (below) and the Palace of Versailles in France, contained mazes. This one at Longleat even has bridges along its paths. You enter the maze, find your way to the center, which is marked, then find your way out. Logic can provide a quicker solution (see above) to reaching the center and finding your way out again.

1 OPEN THE board and turn it over. Tape the pieces of wood to the underside across the fold, so that the board does not snap shut.

2 MARK the center of the maze with paper. Use string to lay a random path through the middle of the checks. Fix it in place with putty adhesive.

3 MAKE WALLS around the string with modeling clay, following the edges of the squares. This route will be the true path through the maze. Then build other walls on the board to create false paths and dead ends.

4 TRY TO make your network quite intricate, covering the whole board. Once the maze is complete, use scissors to cut away parts of the walls on the true path, to create wrong turnings down the dead-end paths.

5 REMOVE the string and putty adhesive from the true path. Make sure that you leave no putty adhesive or greasy marks on the board, so as not to give away the location of the path. Your maze is now ready for use.

6 INVITE A friend to try out the maze. Place a ball at the start. See if your friend can tilt the board to roll the ball through the maze so that the ball reaches the center.

Planning pathways

A FLOW CHART IS A GRAPHIC REPRESENTATION OF A PROCESS. It can be used to plot the stages of any process, from making a cup of tea to designing a piece of software. Flow charts are planned in steps, which are written in boxes, with arrows between the boxes showing the order in which steps are taken. Different box shapes signify data input and output, questions, decisions, and instructions. These charts are used in the design of computer programs and are also an excellent tool for breaking a complex situation, such as the behavior of a species or the traffic flow in a city, into smaller sequences.

▒ Assembling the machine

This chocolate-egg dispenser is made from household items and art materials. A rough design was drawn first, and the main stages written out, then a flow chart was made from this information (right). The machine was assembled and tested by following the chart. As with a logical argument, each part works only if the previous stages work. Try making this machine, or plan your own and build it with a flow chart.

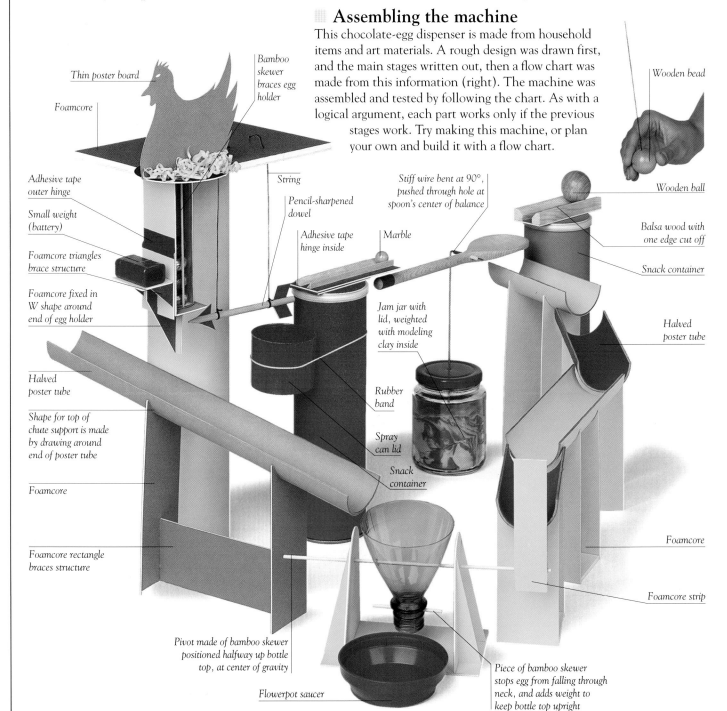

Thin poster board

Foamcore

Bamboo skewer braces egg holder

Wooden bead

Adhesive tape outer hinge

Small weight (battery)

Foamcore triangles brace structure

Foamcore fixed in W shape around end of egg holder

String

Pencil-sharpened dowel

Adhesive tape hinge inside

Marble

Stiff wire bent at 90°, pushed through hole at spoon's center of balance

Wooden ball

Balsa wood with one edge cut off

Snack container

Halved poster tube

Shape for top of chute support is made by drawing around end of poster tube

Foamcore

Foamcore rectangle braces structure

Jam jar with lid, weighted with modeling clay inside

Rubber band

Spray can lid

Snack container

Halved poster tube

Foamcore

Foamcore strip

Pivot made of bamboo skewer positioned halfway up bottle top, at center of gravity

Flowerpot saucer

Piece of bamboo skewer stops egg from falling through neck, and adds weight to keep bottle top upright

Planning and using the chart

Write START at the top of a large sheet of paper. Below, write the stages in boxes as shown here. Ask questions at each stage. If the answer is "no," draw an arrow labeled NO leading to an instruction box, then draw an arrow from the box back to the arrow just before the question. For "yes," draw an arrow, with YES, to an instruction box for the next stage.

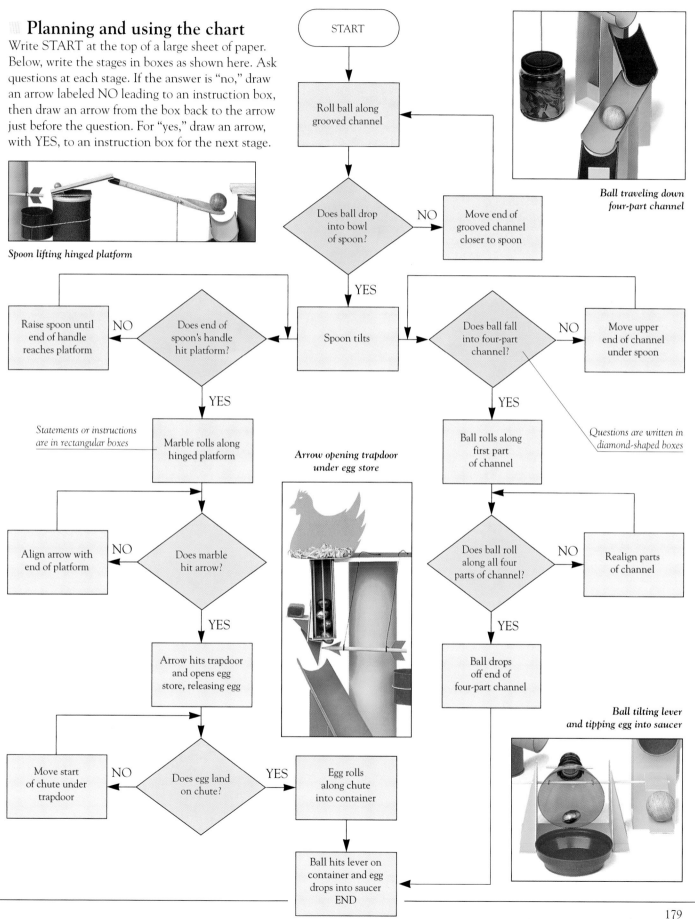

Spoon lifting hinged platform

START

Roll ball along grooved channel

Does ball drop into bowl of spoon? → NO → Move end of grooved channel closer to spoon

YES

Spoon tilts

Does end of spoon's handle hit platform? → NO → Raise spoon until end of handle reaches platform

YES

Statements or instructions are in rectangular boxes

Marble rolls along hinged platform

Does marble hit arrow? → NO → Align arrow with end of platform

YES

Arrow hits trapdoor and opens egg store, releasing egg

Does egg land on chute? → NO → Move start of chute under trapdoor

YES → Egg rolls along chute into container

Does ball fall into four-part channel? → NO → Move upper end of channel under spoon

YES

Questions are written in diamond-shaped boxes

Ball rolls along first part of channel

Does ball roll along all four parts of channel? → NO → Realign parts of channel

YES

Ball drops off end of four-part channel

Ball hits lever on container and egg drops into saucer
END

Ball traveling down four-part channel

Arrow opening trapdoor under egg store

Ball tilting lever and tipping egg into saucer

179

Chaos theory and fractals

IN EVERYDAY LANGUAGE, "chaos" means a state of confusion. In mathematics, however, a chaotic system is one whose behavior over time is hard to predict. This is because the system will behave very differently with only slightly different starting conditions. An example is the weather. If air pressure, temperatures, and wind strength are only slightly affected, then tomorrow's weather may be very different from today's. This is called the "butterfly effect," from the idea that the flap of a butterfly's wing in China could transform weather systems across the world. A fractal is a picture or object that is built up from small shapes. The shape of the overall fractal is similar to the shapes from which it is built. For example, a twig on a tree has a branching pattern similar to that of the whole tree. Fractals can be used to analyze chaotic systems. Chaos theory deals with dynamics, or changes within a system, while fractals show these changes as geometrical images.

■ DISCOVERY ■
Benoit Mandelbrot

Mandelbrot (b. 1924), an American mathematician, developed a branch of mathematics called fractal geometry, which has helped mathematicians to analyze irregularities in systems. He began his study while working out problems for the computer firm IBM. The subjects of his research ranged from fluctuations in the stock market to linguistics and galaxy clusters. Mandelbrot realized that all these subjects were interrelated. They showed patterns of changes that were random, but when these patterns were reduced to smaller elements, the variations kept the same pattern. In 1975 he published *The Fractal Geometry of Nature*, a book of beautiful computer graphics that illustrated the principle of fractal geometry. The Mandelbrot set, which is a pattern made by feeding an equation into a computer repeatedly, was named after him.

Testing chaos

Even simple events can be chaotic: from the same beginning you can have any number of endings. You can see this in the swing of a jointed pendulum.

YOU WILL NEED
- *brass fasteners* ● *nuts*
- *bradawl* ● *foamcore*
- *ruler* ● *pencil* ● *craft knife* ● *cutting mat*

Adult help is advised for this experiment

1 CUT THE FOAMCORE STRIPS to make the two arms of the pendulum. Here, the pendulum arms are 1 in (2.5 cm) wide. One arm is 20¼ in (51 cm) long, and the other arm is two-thirds the length of the first. Ask an adult to help you make a hole in each arm, ½ in (12 mm) from one end.

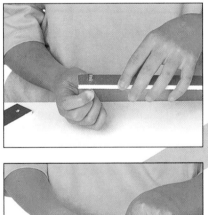

2 PUSH a brass fastener through the end of the longer foamcore arm. Pull the end of the pin so that the head lies against the foamcore, but do not open out the fastener.

3 PUT a nut onto the fastener. This separates the arms so that they swing freely. Then fit the fastener through the other arm and open the ends. Thread a fastener and nut on to the free end of the longer arm.

The pendulum swings
Attach the free brass fastener to a shelf, or to the edge of a piece of wood, by fixing the fastener with adhesive tape. Let the pendulum hang vertically. Push the longer arm to set it swinging, and watch how the shorter arm behaves. For a few moments it will swing wildly around the fastener, then it will calm down and swing backward and forward. Swing the pendulum again, and it will not behave exactly as before. However many times you repeat this action, the movement will be chaotic.

Chaos in business

Buyers in International Financial Futures exchanges (such as the London market, shown here) aim to make a profit by purchasing shares in commodities or currencies that they believe will increase in value. Previously, they relied on estimation, but now chaos theory is used to make these predictions more accurate. The mathematicians who use this theory to advise traders are now some of the most sought-after people in finance.

The making of a snowflake

These color-enhanced images show that every snowflake is unique, yet they all have a six-fold symmetry (p. 158). As a snowflake falls, imperfect crystals stick to it, which builds up the shape in an infinite number of ways. Snowflakes are not true fractals, but a computer can reproduce them like fractals by generating a branched shape and copying it six times. The random nature of this process is of key interest to scientists.

Glossary

Italics indicate a word that has its own entry in the glossary.

ACUTE ANGLE An *angle* that is less than 90°.

ADJACENT In *trigonometry*, the adjacent side of an *angle* in a right-angled *triangle* is the side between the angle concerned and the *right angle*.

ALGEBRA A branch of mathematics in which letters are substituted for *numbers*.

ALGORITHM A step-by-step procedure for solving a mathematical problem. Named after al-Khwarizmi (p. 68), an Arab mathematician.

ANGLE The shape formed by two lines diverging from a point. Angles are measured in degrees; 360° is a full turn.

ARC A section of a *curve* or of a *circle's circumference*.

Bisected line
The slanting black line signifies that the blue line has been bisected, or cut into two equal lengths.

AREA The size of a surface, measured in square *units*. The areas of some shapes can be found using *formulas*.

ARITHMETIC Branch of mathematics concerned with the properties of *numbers*, and with the four basic *operations*.

AVERAGE In *statistics*, the average indicates the typical item in a group of data. It can be found in one of three ways. See *mean*, *median*, and *mode*.

AXIS (plural, AXES) A line used to locate a point on a *graph*. Graphs have a *horizontal* axis, called the *x* axis, and a *vertical* axis (the *y* axis). Each axis has an arrow on the end, and has

numbers or other data along it for plotting points or bars.

BAR CHART A way to show data visually. It has two *axes*, and bars are drawn for each item in a group of data. The length of the bars shows the height, *number*, or other feature of each item.

BASE The *number* of single digits in each *place value* of a *number system*. In base 10, the lowest place value includes the digits 0–9, and 10 is the start of the next place value. In base 2 the lowest place value has just 0 and 1, and 2 (written as 10) is the start of the next.

BINARY A system of counting in *base* 2. Binary *numbers* are made up of a row of 0's and 1's.

BISECT To divide a line or shape into two equal parts.

BRACKETS These separate out small parts of a calculation or *equation*, which are worked out before any other *operation*. In the calculation (3 + 4) × 5, for example, (3+4) is worked out first, and the result put back into the larger calculation. So (3 + 4) × 5 = 7 × 5 = 35.

CARDIOID A closed, heart-shaped *curve* that is the *locus* of a point on the *circumference* of a *circle* as it is rolled around another circle of equal size.

CATENARY A *curve* shaped like a chain hung between two posts or like the thread joining two spokes in a spider's web.

CATENOID A surface with two circular ends, formed by rotating a *catenary* around an imaginary line connecting the central points of the two ends.

CHART A graphic presentation of information so that similar items can be compared, as in a price list, or so that data can be converted from one

system to another (liters to gallons, for instance).

CIRCLE A perfectly round and symmetrical closed *curve* that is the *locus* of all points moving at a constant distance (the *radius*) from a fixed point (the center of the circle).

CIRCUMFERENCE The outside edge of a *circle*. Also the length of that edge.

COMMON FACTOR A *factor* that two or more *numbers* have in common. For example, 30, 60, and 90 have the common factors 1, 2, 3, 5, 6, 10, 15, and 30.

COMPLEMENT To calculate the complement of a *number*, that number is subtracted from the *base*. In base 10, for example, the complement of 7 is 3 (that is 10 – 7).

CONE A solid shape with a circle at the base and a curved surface rising to a point. Cut horizontally, it has a circular *cross-section*. Cut in half vertically, it has a *triangular* cross-section. Cut at an angle, it shows an *ellipse*.

CONVERSION The transformation of data from one form or system of measurement into another; for example, from Centigrade into Fahrenheit.

COORDINATES Pairs of *numbers* that define the location of

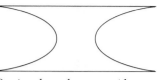

Section through a catenoid
A section lengthwise through the center shows the catenary curves that form the surface between the ends.

a point relative to another point or an *axis*. Cartesian coordinates (p. 74) are always written with the *x* axis (horizontal axis) first.

COSINE A trigonometric *ratio* (often shortened to cos). To find the cosine of an angle in a right-angled *triangle*, divide the length of the *adjacent* side by that of the *hypotenuse*.

CROSS-SECTION A shape made by cutting a solid object along a line usually at *right angles* to its height or length.

CUBE A solid shape with six identical *faces*, all *squares*.

CUBOID A solid shape with six *rectangular faces* that may (but need not) be *square*.

CURVE The *locus* of a point moving in specified conditions to form a continuous bend. Open curves have end points, while closed curves have joined ends.

CYCLOID The *locus* of a point on the *circumference* of a *circle* that rolls along a straight line. It resembles a flattened arch.

Parts of a circle
A chord divides a circle into two segments. A diameter bisects it into two semicircles (equal segments). A sector is a slice of the circle; each straight edge is a radius, and the curve is an arc (part of the circumference).

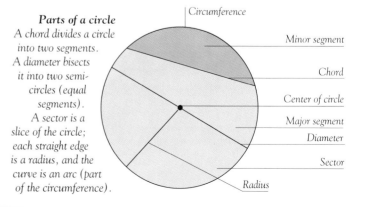

Circumference
Minor segment
Chord
Center of circle
Major segment
Diameter
Sector
Radius

CYLINDER A solid shape with a *curved* surface joining the edges of two identical *circles* or *ellipses*. A circle-based cylinder cut parallel to the base has a circular *cross-section*, but if cut lengthwise it would show a *square* or *rectangular* section.

DECIMAL A *number* expressed using the decimal counting system, in hundreds, tens, *units*.

DECIMAL PLACE In a *decimal*, the place that a *digit* occupies in relation to the *decimal point*.

DECIMAL POINT The point in a *decimal*. *Digits* with a value of 1 or more lie to the left of the point, and those less than 1 lie to the right.

DENARY SYSTEM Another name for *base 10* or the *decimal* counting system.

DIAGONAL A straight line between any two non-adjacent corners of a *polygon*, or joining one corner of a *polyhedron* to another not in the same *plane*.

DIAMETER A line that exactly bisects a *circle*. It starts and ends on the *circumference*, and passes through the center. It has twice the length of the *radius*.

DIGIT A *number* represented by a single *numeral*. The digits used in the Western *number system* are 0 1 2 3 4 5 6 7 8 9.

DIMENSION A way of defining space in terms of the way it can be measured. Points in one dimension lie on a straight line or a *curve*, which has only length; those in two dimensions (2D) lie on a flat surface, which has length and breadth; those in three dimensions (3D) are located within a volume, which has length, breadth, and height.

DIRECTRIX A straight line used as a basis for plotting a shape, such as a *parabola*.

DIVINE PROPORTION Another name for the *Golden Section*.

DIVISOR A *number* that is used to divide another number.

ELEVATION On solid shapes, such as buildings, an elevation is the view you see if you look at one *face* of the shape.

ELLIPSE A closed *curve* that looks like an elongated *circle*. It has two *lines of symmetry*. It also has two *foci*.

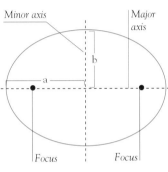

Finding the area of an ellipse
This is calculated using the formula πab, where a is half the length of the major (long) axis and b is half the length of the minor (short) axis.

EQUATION A mathematical statement consisting of two pieces of information separated by an = sign; the information on one side of the sign is equal to that on the other side.

EQUIVALENT FRACTIONS *Fractions* that are expressed in different ways but have the same value, such as ½ and ¼.

ESTIMATION The art of making an educated guess. It is used to give a rough idea of a measurement or an answer, without full calculations.

EXPONENT The *power* to which a *number* or *algebraic* letter has been raised.

FACE A flat surface on a three-dimensional shape.

FACTOR The factors of a particular *number* are all the numbers that divide into it exactly. For example, the factors of 6 are 1, 2, 3, and 6.

FOCUS (plural, FOCI) A fixed point used as the basis for plotting an *ellipse* or a *parabola*.

FORMULA A general rule that is followed in order to obtain a particular result.

FRACTION A *ratio* of two *integers*. It is shown as two *numbers* separated by a line — for example, ¾. The number above is called the numerator. It shows how many parts there are in the fraction. The one below — the denominator — represents the number of parts making up the whole.

GEOMETRY The study of space — points, lines, flat shapes, and solid shapes.

GOLDEN SECTION This is a division of a line into two sections so that the *ratio* of the larger part to the smaller is the same as that of the whole line to the larger part. This ratio is roughly 1.618:1.

GRAPH A visual illustration of data. It usually has two *axes*, and has points plotted on it. These points may be joined with a line, or left as dots (as in a *scatter diagram*).

HELIX (plural, HELICES) An open *curve* around a fixed line (*axis*), which forms a three-dimensional *spiral*.

HIGHEST COMMON FACTOR (HCF) The largest of all the *factors* common to two or more *numbers*. For example, the HCF of 45 and 75 is 15.

HORIZONTAL A horizontal line or surface is one that lies flat, at an angle of 0° or 180°.

HYPOTENUSE The longest side of a right-angled *triangle*, found opposite the *right angle*.

IMPERIAL SYSTEM A system of weights and measures including ounces, inches, and gallons.

INDEX A small figure written above and to the right of a *number* or *algebraic* letter, indicating the *exponent*. For example, in 2^3 (2 to the power of 3), 3 is the index.

INTEGER A whole *number*, such as 1, 2, 3, 0, –1, –2.

IRRATIONAL NUMBER A *number* that cannot be expressed exactly as either a *fraction* or a *decimal* number. One notable example is *pi*.

LINE OF BEST FIT A straight or *curved* line on a *scatter diagram* that passes through the region where the points are most thickly clustered.

LINE OF SYMMETRY A line dividing a shape into two pieces that are exactly symmetrical (see *symmetry*).

LOCUS (plural, LOCI) In *geometry*, a set of points that all meet certain conditions. For example, the locus of all points at a given distance from a point on a *plane* is a *circle*.

Locus of points around a square
The dotted line, with quarter-circles at the corners, is the locus of all points ½ in (1 cm) outside the edges.

LOWEST COMMON MULTIPLE (LCM) In a group of *integers*, this is the smallest *number* that can be divided exactly by each integer. For example, the LCM of 3 and 5 is 15.

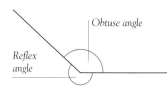

Obtuse and reflex angles
An obtuse angle is an angle that has a value between 90° and 180°. A reflex angle is greater than 180° but less than 360°.

MEAN In a group of data, the mean is an *average* taken from the whole group. To calculate it, add together the values of every item in the group, and divide the result by the number of items in the group.

MEDIAN The item that is in the middle of a group of data. For example, among the *numbers* 1, 4, 5, 7, 8, 9, 10, the median is 7. In even-numbered groups with no single middle number, the median is the *mean* of the two middle numbers.

METRIC SYSTEM A system of measurement based on *units* that can be divided exactly by 10, 100, or 1,000.

MODE In a group of data, the mode is the type of item that occurs with the greatest frequency.

MULTIPLE The multiples of an *integer* are all *numbers* created by multiplying that integer by another. Times tables show multiples of 1, 2, 3, and so on.

NEGATIVE NUMBER A *number* less than 0. It is written with a minus sign directly in front of it: for example, –3.

NET In *geometry*, a flat shape that can be folded and have its edges joined together to make a solid shape.

NUMBER An expression that denotes an amount. Numbers are expressed as combinations of *digits*. They may be *decimals*, or may be written as *fractions*.

NUMBER SYSTEM A method of writing *numbers* using a set of symbols, or *numerals*. Systems in use today include the Western, Roman, Arabic, and Chinese systems.

NUMERAL A symbol that represents a *digit*.

OBTUSE ANGLE An *angle* of between 90° and 180°.

OPERATION An *arithmetical* process applied to two or more *numbers* or *algebraic* terms. The most basic operations are addition, subtraction, multiplication, and division.

OPPOSITE In *trigonometry*, the side of a right-angled *triangle* opposite the *angle* being measured.

PARABOLA An open *curve* that is plotted from a *focus* and a *directrix*, and *bisected* exactly by an *axis*. It is the *locus* of all points for which the distance from that point to the focus is the same as the distance from that point to the directrix.

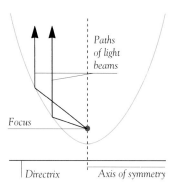

Parabola
This symmetrical curve is used as the basic shape for the insides of headlamps. Light shining from the focus hits the parabola's edges and is reflected outward in parallel lines.

PARALLEL Lines that are parallel lie side by side, and preserve the same distance between them along their entire length. Pairs or groups of parallel lines can be either straight or *curved*.

PARALLELOGRAM A flat, *quadrilateral* shape whose opposite sides are *parallel* and of equal length, and whose opposite *angles* are equal. Parallelograms include *squares*, *rectangles*, and *rhombi*.

PERCENTAGE A way of expressing an amount as a part of 100. For example, 37% means 37 out of 100.

PERFECT NUMBER A *number* that is the sum of all its *divisors* except the number itself. An example is 6, which is the sum of $1 + 2 + 3$.

PI Written as π, this is the *ratio* between the *circumference* and the *diameter* of a *circle*. As an *irrational number* it is impossible to define exactly; it has been calculated to hundreds of *decimal places*, but is commonly given as 3.142.

PIE CHART A way of showing data visually, with each item of data occupying a sector (p. 182) in a *circle*.

PLACE VALUE A value given to a *digit* in a *number* by its position in that number. The right-hand digit in the number has the lowest place value, and the left-hand digit has the highest. Place values increase in *powers* of 10 or whichever other number is being used as the *base*. In the *decimal* system, the lowest place value of a whole number is a unit (10^0), followed by a ten (10^1), then a hundred (10^2), and so on.

PLANE A flat surface. A plane shape is a flat shape.

PLATONIC SOLIDS *Polyhedra* with *faces* formed from one type of regular *polygon*.

POLYGON A flat shape with three or more straight sides.

POLYHEDRON A solid shape with four or more *faces*.

POSITIVE NUMBER A *number* greater than 0. It may be written with or without a + sign in front of it.

POWER Denotes the number of times by which a *number* has been multiplied by itself. The power is indicated by the *exponent*. For example, 2^3 is $2 \times 2 \times 2$, or "2 to the power of 3."

Quadrilateral
This type of shape can take any form, but all quadrilaterals have four sides. The internal angles always add up to 360°.

PRIME NUMBER A *number* that can be exactly divided only by itself and by 1. (The number 1 is not a prime number.)

PRISM A three-dimensional shape whose *cross-sections*, cut parallel to an end *face*, are the same shape as the end faces.

PROBABILITY The likelihood of an event happening. It can be written as *numbers* separated by an oblique line (1/10), as a *decimal*, or as a *percentage*.

PROJECTION Drawing a three-dimensional shape, such as a building, in a *plane*.

PROPORTION The relationship between *numbers* in a *ratio*.

QUADRILATERAL A two-dimensional shape that has four straight sides.

RADIUS The distance from the central point of a *circle* to the *circumference*.

RATIO A way of showing a comparison between amounts. The amounts are usually written as *numbers* separated by a colon (for example, 1:4) or by an oblique line (1/4). These expressions mean "one part to four," and are not to be confused with a *fraction*.

RECTANGLE A *quadrilateral* shape in which opposite sides are *parallel* and of equal length, and all of the internal *angles* are *right angles*.

RECURRING DECIMAL A *decimal number* which, after a particular point, has an endlessly repeated *digit* or set of digits. A dot above the start and end shows the repetition – so 0.4̇1̇5 = 0.415415415 . . .

REFLECTION A reproduction of a shape on the other side of a line that lies on or outside that shape. The reflection has exactly the same form as the original shape but is reversed.

REFLEX ANGLE An *angle* greater than 180°.

RHOMBUS (plural, RHOMBI) A *parallelogram* whose sides are all of equal length. The shape also has opposite pairs of equal *angles*.

RIGHT ANGLE An *angle* of exactly 90° (a quarter-turn).

ROTATE To turn a shape through a certain *angle*, around a given point called the center of rotation. If a shape is rotated about its central point, and looks the same when it is in its new position, it is said to have rotational *symmetry*.

ROUNDING OFF Reducing a *number* to a reasonable size. Numbers are usually rounded off to one or two *decimal places* (d.p.) after the *decimal point*. To round off to 2 d.p., look at the *digit* in the third decimal place. If it is 4 or less, cross off the digits after it; if it is 5 or more, add 1 to the digit to the left of it, then cross off the rest.

SCATTER DIAGRAM A *graph* on which points are plotted to show a combination of two features, such as the ages and the shoe sizes of a group of children. A *line of best fit* may be drawn on the diagram to show the general relationship between the two features.

SIGNIFICANT FIGURES The number of *digits* to which a *number* is reduced when *rounding off*. For example, reducing 3.14159 to four significant figures gives 3.142.

SINE A *trigonometric ratio*. On a right-angled *triangle*, the sine of a particular *angle* is found by dividing the length of the *opposite* side by the length of the *hypotenuse*.

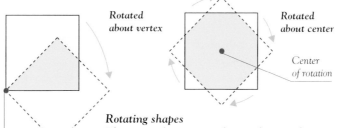

Rotated about vertex

Rotated about center

Center of rotation

Center of rotation

Rotating shapes
The center of rotation can lie anywhere in relation to a shape — outside it, at one edge, or in the center.

SPHERE A round, regular, three-dimensional shape whose surface is the *locus* of all points that lie at a given distance from the center.

SPIRAL An open-ended *curve* that coils around a given point or *axis*. It can be either flat or three-dimensional (see *helix*).

SQUARE (1) A *rectangle* whose sides are all of equal length. (2) Short for *square number*.

SQUARE NUMBER A total that is produced by multiplying a *number* by itself once: for example, 81 = 9×9.

SQUARE ROOT The *number* that is multiplied by itself to produce a *square number*; the square root of 81 is 9.

STATISTICS *Numbers* derived from, or making up, items of data. They are processed in particular ways to allow specific conclusions to be drawn from the information.

SYMMETRY A two- or three-dimensional shape has symmetry if it looks the same when transformed in certain ways (e.g. rotated through a particular angle or reflected on the other side of a line).

TABLE A way of arranging data in rows and columns, to show the relationships between the items.

TANGENT (1) A line that touches the edge of a *circle* or a *curve* at one point. (2) A *trigonometric ratio*. In a right-angled *triangle*, the tangent of a given *angle* is calculated by dividing the length of the *opposite* side by the length of the *adjacent* side.

TESSELLATION A pattern made from flat shapes that are repeated and fitted together exactly, so that no space is left in between them.

TRIANGLE A three-sided *polygon*. Triangles may be equilateral (having sides of equal length), isosceles (with two sides of equal length and two equal angles), or scalene (in which all of the sides and angles are different). Some isosceles and scalene triangles are right-angled (having one internal angle that is a *right angle*).

TRIGONOMETRY The study of *ratios* called "trigonometric functions" and the features of *triangles*. It is usually concerned with right-angled triangles.

UNIT (1) The lowest positive *integer*: 1. (2) The *number* immediately to the left of a *decimal point*. (3) A basis for measurement; for example, a foot (or a meter) is a unit of length, and a pound (or a kilogram) is a unit of mass.

VERTEX (plural, VERTICES) A point on a flat shape where two or more lines meet, or a point on a solid shape where two or more *faces* meet.

VERTICAL A word describing something that is upright.

Vertices, edges, and faces
A vertex is a corner on a flat or solid shape. A flat shape (right) has as many vertices as edges, and has only one face. For solid shapes (far right), the relationship between the number of vertices, edges, and faces is expressed by Euler's theorem (p. 164).

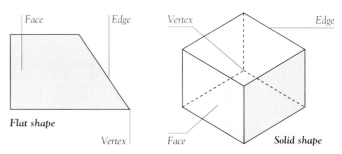

Face   Edge

Flat shape   Vertex

Vertex   Edge

Face   **Solid shape**

Answers to problems

p. 17 Gray matter (top)
1) c   (2) b   (3) c

p. 17 Gray matter (bottom)
$35 + 64 = 99$ $22 + 41 = 63$
$60 \div 15 = 4$ $19 \times 3 = 57$
$75 - 60 = 15$ $121 \div 11 = 11$
$999 - 333 = 666$ $7 \times 63 = 441$

p. 20 Puzzle
1274953680 is wholly divisible by all the numbers from 1 to 16. It also has all the digits from 0 to 9.

p. 21 Permutations
A book of 8 pages split into three would give 512 ($8 \times 8 \times 8$) different permutations.

p. 22 What are these?
The left-hand picture is of empty wine bottles. No magnification was used. The image on the right has been magnified 37 times and enhanced with color. It shows the bristles on a toothbrush.

p. 23 Making Napier's bones
$1572 \times 3 = 4716$

p. 25 Puzzle
Eleven thousand plus eleven hundred plus eleven actually equals 12,111
($11,000 + 1100 + 11 = 12,111$)

p. 26 Puzzle (bottom)
The numbers in the complete magic square are:

7	6	11
12	8	4
5	10	9

p. 27 Puzzle
Each row, column, and diagonal adds up to 260. The numbers in each corner plus the four numbers in the center also add up to 260. The totals for each side of the small square are: 58; 74; 202; 186. $186 + 74 = 260$; $202 + 58 = 260$.

p. 28 Acceleration
Acceleration will be roughly the same, whatever the distance.

p. 29 Making a pulley
With two pulleys the effort equals roughly half the weight of the load, but you will pull about twice the distance.

p. 31 Making a thermometer
The level in the straw should drop when the thermometer is placed in the ice. The liquid in the straw contracts when the temperature is low, and expands when it is hotter.

p. 32 Gray matter
$135 - 24 = 111$
$214 + 5 + 3 = 222$
$345 - 12 = 333$

p. 34 Puzzle
The decimal square is magic, with each column, row, and diagonal adding up to 3.75. The numbers are:

2.00	0.25	1.50
0.75	1.25	1.75
1.00	2.25	0.50

p. 40 Odd numbers and powers
The sum of the first n odd numbers is n^2; for example, $1 + 3 + 5 + 7 + 9 + 11 = 36$ or 6^2.

p. 40 Puzzle
$25 = 5^2$; $100 = 10^2$; $144 = 12^2$; $169 = 13^2$

p. 41 Puzzle
Squares for numbers with 1 are:
$$1^2 = 1$$
$$11^2 = 121$$
$$111^2 = 12321$$
$$1111^2 = 1234321$$
$$11111^2 = 123454321$$
$$111111^2 = 12345654321$$
$$1111111^2 = 1234567654321$$

Squares for numbers with 3 are:
$$3^2 = 9$$
$$33^2 = 1089$$
$$333^2 = 110889$$
$$3333^2 = 11108889$$
$$33333^2 = 1111088889$$
$$333333^2 = 111110888889$$
$$3333333^2 = 11111108888889$$

p. 43 Gray matter
Even as you read this, you are adding on seconds, so round off to the nearest day. As an estimate, if you are 10 years old, you should have a figure of around 300 million seconds.

p. 44 Gray matter
The next five numbers in the sequence are: 73, 89, 107, 127, 149. They are all prime numbers. This formula does not work for primes less than 17, or for values of n greater than 15.

p. 49 Sissa's reward
Sissa was very clever. By doubling the amount on each successive square, he would have amassed 4,000 million grains of rice by the time he was only halfway down the chess board.

p. 55 Population density
Hong Kong has the highest population density of any country in Southeast Asia: 13,223 people per square mile (5,106 per square kilometer). This is more than nine times as dense as the next most populous country, Taiwan, which has 1,418 people per square mile (547 per square kilometer).

p. 62 Floating ice (see below)
When the ice cubes melt, the water level in the glass will stay the same. Because water expands when it freezes, an ice cube is slightly larger and less dense than the water from which it is made. When floating, it displaces a volume of water that is equal to the volume of water from which it is made.

Glass showing water level after ice cubes have melted

p. 62 Puzzle
6729/13458   7932/15864
9327/18654   7923/15846
6927/13854   7329/14658

p. 63 Freezing water
When water freezes, it increases in volume by about one-sixth, so the water when frozen should reach the next mark.

p. 63 How much land is there?
Approximately one-third of the Earth is covered by land. You will need to simplify the fraction greatly to get this estimate.

p. 70 Puzzle
Let M = the mother's age and f = the friend's age. You could show the mother's present age as 3f = M. In 15 years' time the mother would be twice as old as her son, so the equation for her age would be $M + 15 = 2(f + 15)$, giving $2f = M - 15$. Subtract the second equation from the first as shown:
$$3f = M$$
$$\underline{2f = M - 15}$$
$$f = 0 \ + 15$$

p. 70 Showing patterns
Each new square needs 3 pencils; the first square had 3 pencils + 1 to start the line. Therefore, the formula is p = 3S + 1.
For 10 squares you need (3×10) + 1 or 31 pencils; for 300, (3×300) + 1 or 901 pencils.

p. 71 Trick
The answer is x.

p. 73 Function machine
The function applied in each case is $f(x) = 2x + 1$. Applied to 3, 4, and 5, this gives:
$(2 \times 3) + 1 = 7$
$(2 \times 4) + 1 = 9$
$(2 \times 5) + 1 = 11$

p. 73 Puzzle
Let a = the person's age.
$$a = 3(a + 3) - 3(a - 3)$$
$$= 3a + 9 - 3a + 9*$$
$$= 3a - 3a + 9 + 9$$
$$= 9 + 9$$
$$= 18$$
Check: $a = (21 \times 3) - (15 \times 3)$
$$= 63 - 45$$
$$= 18$$
* minus × minus = plus

p. 74 Coffee temperatures
The coffee with the milk put in first should keep warm longer than the coffee with the milk added after 5 minutes.

p. 80 Puzzle
If the black ball is taken out, you cannot be sure if the ball remaining inside the bag is an original ball (which could be white or black) or the new one (which is black). This gives three possibilities:
black, black, white
so the probability of taking out black is 2/3;
the probability of taking out white is 1/3.
If a white ball is taken out, this must be original, so black (new) is left inside the bag. The probability that the remaining ball is black will therefore be 1, and the probability that the ball is white will be 0.

p. 93 Puzzle
Three times.

p. 96 How far away is the Moon?
The ratio of the counter-eye distance to the counter's diameter should be about 111:1. If the Moon's diameter is 2,160 miles (3,475 km), the distance from the Moon to the Earth is 2,160 miles (or 3,475 km)×111 or about 238,857 miles (384,320 km).

p. 98 Puzzle
See below.

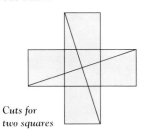

Cuts for two squares

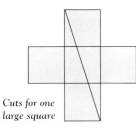

Cuts for one large square

p. 98 Same perimeter
Using the same perimeter, the circle will give the greatest area.

p. 103 Experimenting with density
The small piece of modeling clay will have the same density as the larger piece. The density of a particular material remains the same no matter what the volume of that material.

p 106 Puzzle (left)
The volume of 2.2 lbs of water is one quart.

p. 106 Puzzle (right)
Cream is less dense than milk, even though it appears to be thicker. This is why the cream rises to the top whenever the two liquids are mixed.

p. 109 Swinging pendulums
If you reduce the length of the pendulum, the bob will take less time for each swing. If you increase the length of the pendulum, each of the swings will take a longer time than before.

p. 121 Puzzle
The coin that fits will be smaller than you think. The shearing of a shape creates an optical illusion.

p. 122 Puzzle
Lay one plank to form a chord (p. 182) across the outer bank of the bend. Lay the other from the center of the first plank to the inner bank. This makes a right-angled triangle with straight sides of 9 ft and 4 ft 6 in (2.75 m and 1.37 m). Pythagoras' theorem (p. 124) gives $9^2+4.5^2 = 101.25$. The square root of 101.25 is just over 10, so a 9 ft plank will just reach across the moat.

p. 122 Sum of angles of triangle
The internal angles of a triangle add up to 180°. (See p. 161 for an exception to this rule.)

p. 123 Puzzle
There are 12 in all; count the stacked triangles and the ones covering more than one section.

p. 123 Strength in triangles
The triangle cannot be distorted. If you use four straws, the shape will shear to a parallelogram.

p. 129 Quadrilateral angles
All internal angles of a quadrilateral add up to 360°.

p. 130 Tangrams
See below.

Figure with bowl

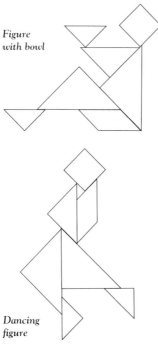

Dancing figure

Sitting cat

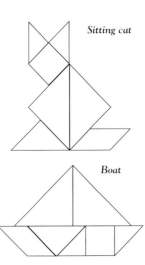

Boat

p. 135 Count Buffon's estimation of π
Suppose you throw the match 30 times and it hits the line 18 times. Divide the number of checks by the times the match was thrown. The experiment gives ¹⁸⁄₃₀ = 0.6. This is close to ²⁄π, which is 0.6365.

p. 148 Spring balance
The spring extends twice as much for twice the load.

p. 160 Traveling in a straight line
The route will actually form a curved line on the map.

p. 161 The sum of the angles
The angles of the convex triangle total slightly more than 180°. Those of the concave triangle add up to slightly less than 180°.

p. 165 Cutting a Möbius strip
A line drawn along the center of the strip will join up with itself. A strip cut in two forms a doubly long strip with two twists in it; if cut in two again, it forms two linked strips with double twists. A strip cut in thirds forms another strip linked to a doubly long loop.

p. 165 Trick
Hold your rope at the middle. Make a small loop, and pass it under the rope on your friend's right wrist, pushing it toward his fingers. Pass the loop over his hand, and work it back under the rope on his wrist. You should now be unlinked, although each of you will still have your wrists tied.

p. 166 Puzzle
The minimum number of colors needed is 2. On a geographical map, the minimum number needed so that countries of the same color do not share a boundary is 4.

p. 172 Think about it
If you jump in the air while on a train, you will still move forward at the same speed as the train.

p. 174 Puzzle
See diagram below.

p. 175 Puzzle
The only shape whose edges you can trace without going over any line twice is the octahedron.

p. 175 Tower of Hanoi
The minimum number of moves needed to transfer all the rings in order is 2^n-1, where n is the number of rings. For example, to transfer five rings you would need to make at least 31 moves.

p. 176 Puzzle
There will be the same amount of red-colored water in the green glassful as there is green water in the red glassful.

p. 176 The tale of the journey across the stream
There are at least three possible answers for this puzzle. One is as follows. Take hen across. Return with empty boat. Take fox. Bring hen back. Take grain. Return with empty boat. Take hen across again.

Index

Acknowledgments

CAROL VORDERMAN would like to thank: everyone at Dorling Kindersley who has been involved with the project for their patience, creativity, and total professionalism, Jim Miller for his help in checking the book, my family — Patrick, Katie, and Jean — for keeping the house quiet when I needed to write, and my mathematics teacher Mr. Parry for his inspiration when I was young.

DORLING KINDERSLEY would like to thank: The Rev. Dr. Paddy FitzPatrick, for his material on calendars and time measurement; Graeme Hill and George the dog; Hampstead Garden Centre; R.D.H. Walker, Junior Bursar, Queens' College, Cambridge, England, for information on the "mathematical bridge"; Wanda G. Xu, for explaining the use of the Chinese abacus.

Thanks also to Josie Buchanan, for initial planning work on the book; Sarah Angliss, for editorial help; Stephanie Jackson; Jane Parker, for the index; Stephen Stuart, for production. Picture research by Deborah Pownall.

Photography by Andy Crawford. Photographic assistance from Gary Ombler, Pauline Naylor, and Nick Goodall. Additional photography by: Tina Chambers, Geoff Dann, Phill Dowell, Steve Gorton, Frank Greenaway, Peter Hayman, Chas Howson, Colin Keates, Dave King, Bob Langrish, Karl Shone, Clive Streeter, Andreas Von Einsiedel, and Jerry Young.

PICTURE CREDITS

t=top; c=centre; b=bottom;
r=right; l=left.

Ace Photo Agency: 88–89
AKG: 156tr, 162tr
Arcaid: 50–51, 58cr, 118tr, /Michael Bryant 142tr
British Museum: 67t, 106c
By permission of the Syndics of Cambridge University: 40tr
Camera Press: 149tr
Jean-Loup Charmet: 68c, 69bl, 70tr, 74cr, 78cr, 124tl
Bruce Coleman: 30tr
© 1995 M.C. Escher/Cordon Art-Baarn-Holland. All rights reserved. 130cr
© British Crown Copyright/MOD. Reproduced with the permission of the Controller of Her Britannic Majesty's Stationery Office: 124cl
C.M. Dixon: 14tr, 78bl, 94cl
e.t. Archive: 80tr, 99tr
Mary Evans Picture Library: 13tl, 15bl, 29tr, 38bl, 44tr, 52bc, 68br, 71cl, 72tr, 79tl, 93cl, 102tr, 104tr, 116tr, 126tr, 170cl, 173tr, 176tr
FLPA: 148tr
Robert Harding Picture Library: 28tr, 63cl, 115tr, 170cr
Michael Holford: 16cl, 68tl, 90cl, 114tl, 160cr
Angelo Hornak: 177bl
Hulton Deutsch: 18cr, 24tr, 58c, 87tr, 106cr
Image Bank/Terje Rakke: 122tl, 139tr
Image Select/Ann Ronan: 23cr, 60tr
Images Colour Library: 169t
A.F. Kersting: 82cb
Bob Langrish: 92tr
Mansell Collection: 13bl, 33tl, 38cr, 47bl, 114cr, 150br, 164cr
NASA: 31tr, 103cr
National Maritime Museum: 12cr, 20tr, 23br, 108tr
Pictor International: 62tr, 134tr
Range Pictures/Bettmann: 170cr
Rex Features: 36tr
Scala: 11tr, 52cl
Science Museum Library: 102cr, 165tl
Science Photo Library/Andrew Syred: 22tr, /James King Holmes: 34tr, /Andrew Syred: 39cr, /NASA: 42ct, /Allan Morton: 56tr, /Philippe Plailly: 101tr, /Joe Pasieka: 126bc, /Martin Dohrn: 136tr, /NOAO: 146tr, 151tl, /Jerry Mason: 151cr, /Gregory Sams: 168–169, /Hank Morgan: 180bl, /Jerome Yeats: 181tr, /Scott Camazine: 181br
Frank Spooner Pictures: 78cl, 144tr
Tony Stone Images: 57br, 76–77
Telegraph Colour Library/Planet Earth: 10–11
Vatican Library, Pal. Lat.: 1564, 98tr
Zefa: 13cr, 22cr, 33tl, 55tr, 66–67, 108, 109tr, 112–113, 120tr, 126cl, 131cl, 141tr, 150cl

ILLUSTRATIONS

Andrew MacDonald: 29br, 80tr, 157br
Mark Franklin: 53tl, 126br
Geoff Denney Associates: 41t, 107b
Gary Small: 21tl
Phil Ormerod: 69cr, 78cr, 182–185
Paul Carpenter: 79cb, 85tr, 115cl, 117tr, 123br, 128bl, 130br, 131tr, 135br, 152cr, 154cr, 159tr, 162br, 164tr, 175tl, 187bl, 187cb
Elaine Monaghan: 123tl, 166tr
Gurinder Purewall: 129cr

MODEL MAKERS

Peter Griffiths: 2tr, 5ct, 6tr, 12ct, 19r, 61b, 123r, 139br, 142br, 145br, 149br, 157b, 161br, 171tc, 178–179
Rob Newman: 163bl, 167bl
All other models made by Gurinder Purewall, Elaine Monaghan, Katie John

MODELS

Bissy Adejaré; Wolé Adejaré; Tom Baxter; Leon Brown; Sally Brown; Imogen Bryan; Rebecca Bunting; Eleasha Burrell; William Burton; Peter Cao; Jonathan Chen; Charlie Clark; Laura Douglas; Maya Foa; Charlene Fovargue; Miguel Garrido; Emily Gorton; Elizabeth Haher; Jason Haniff; Harry Heard; Sophie Holbrook; Katie John; Charyn Jones; Grace Jones; Jake Jones; Natalie King; Olivia King; Eleanor Lake; Sophie Lindblom-Smith; Lynette Marshall; Courtney McGibbon; Michaela Mitchell; Brian Monaghan; Elaine Monaghan; Humerah Mughal; Vikram Murthy; Pauline Naylor; Michelle Papadopoulos; Priya Patel; Alastair Raitt; Rebecca Shilling; Henrietta Short; Danny Slingsby; Natalie Slingsby; Matthew Smedley; Clive Sun; Andrew Thomas; Naomi Vernon; Ella Walsh; Paul West; Michael Winkless